T0238818

Philip Husbands   Jean-Arcady Meyer (Eds.)

Evolutionary Robotics

First European Workshop, EvoRobot98
Paris, France, April 16-17, 1998
Proceedings

Springer

Series Editors

Gerhard Goos, Karlsruhe University, Germany
Juris Hartmanis, Cornell University, NY, USA
Jan van Leeuwen, Utrecht University, The Netherlands

Volume Editors

Philip Husbands
COGS, University of Sussex
Falmer, Brighton, BN1 9QH, England
E-mail: philh@cogs.susx.ac.uk

Jean-Arcady Meyer
AnimatLab, Ecole Normale Superieure
47, rue d'Ulm, F-75230 Paris Cedex 05, France
E-mail: meyer@wotan.ens.fr

Cataloging-in-Publication data applied for

Die Deutsche Bibliothek - CIP-Einheitsaufnahme

Evolutionary robotics : first European workshop / EvoRobot 98,
Paris, France, April 1998. Philip Husbands ; Jean-Arcady Meyer
(ed.). - Berlin ; Heidelberg ; New York ; Barcelona ; Budapest ; Hong
Kong ; London ; Milan ; Paris ; Singapore ; Tokyo : Springer, 1998
   (Lecture notes in computer science ; Vol. 1468)
   ISBN 3-540-64957-3

CR Subject Classification (1991): I.2.9, F.1.1, I.2, D.1, F.1

ISSN 0302-9743
ISBN 3-540-64957-3 Springer-Verlag Berlin Heidelberg New York

© Springer-Verlag Berlin Heidelberg 1998
Printed in Germany

Typesetting: Camera-ready by author
SPIN 10638562   06/3142 – 5 4 3 2 1 0   Printed on acid-free paper

Lecture Notes in Computer Science 1468

Edited by G. Goos, J. Hartmanis and J. van Leeuwen

Springer

Berlin
Heidelberg
New York
Barcelona
Budapest
Hong Kong
London
Milan
Paris
Singapore
Tokyo

Preface

This volume contains the papers accepted for presentation at the First European Workshop on Evolutionary Robotics (EvoRobot98) held at Ministre de l'Education Nationale de l'Enseignement Superieur et de la Recherche, Paris, 16–17 April, 1998. The workshop was organised by EvoRob, the working group on evolutionary robotics of EvoNet, the European Network of Excellence in Evolutionary Computing. The workshop was supported by EvoNet and brought together most of Europe's leading researchers in the area with some welcome participation from Japan. Papers were refereed by a committee composed of the following people:

Wolfgang Banzhaf, Universität Dortmund, Germany
Marco Colombetti, Politecnico di Milano, Italy
Luis Correia, Universidade Nova de Lisboa, Portugal
Jean-Yves Donnart, ENS, Paris, France
Marco Dorigo, Université Libre de Bruxelles, Belgium
Alexis Drogoul, Université Paris 6, France
Dario Floreano, EPFL, Switzerland
Jelena Godjeva, EPFL, Switzerland
Frederic Gruau, CWI, The Netherlands
Inman Harvey, University of Sussex, UK
Nick Jakobi, University of Sussex, UK
Jerome Kodjabachian, ENS, Paris, France
Henrik Lund, University of Aarhus, Denmark
Francesco Mondada, EPFL, Switzerland
Stefano Nolfi, National Research Council, Italy
Domenico Parisi, National Research Council, Italy
Rolf Pfeifer, University of Zurich, Switzerland
Marc Schoenauer, Ecole Polytechnique, France
Adrian Thompson, University of Sussex, UK
Mike Wheeler, University of Oxford, UK

Many thanks to all of them. Thanks are also due to Jennifer Willies and Julian Miller for their invaluable help with the event and to Marc Schoenauer for excellent local arrangements.

Evolutionary Robotics (ER) is a new field of research that is concerned with the use of evolutionary computing techniques for the automatic design of adaptive robots. The aims of this workshop, were to assess the current state of the art and to provide opportunities for fostering future developments and applications.

The EvoRob working group identified the following topics as being of particular importance to the field:

– Genotype to phenotype mappings
– Analysis of evolutionary processes and of evolved robots
– Adaptation to changing environments
– Co-evolution of control architectures and body plans
– Interactions of evolution, development and learning
– Evolvable hardware in robotics
– Fundamental methodological issues
– Simulation-reality transference
– Comparison of evolutionary robotics methodologies
– Scaling to complex behaviours
– Fitness formula analysis
– Role of evolutionary robotics in cognitive science
– Neuroethological uses of evolutionary robotics

Many of these themes recur throughout the papers collected here. Because the topic of the workshop was already focused, the papers did not divide up into obvious and neat sections, hence they are found here in the order in which they were presented at the workshop. We hope that you will find this volume a useful snapshot of the state of the art of a young field that continues to develop and generate important new ideas.

February 1998   Phil Husbands and Jean-Arcady Meyer

Table of Contents

Evolutionary Robotics: A Survey of Applications and Problems
Jean-Arcady Meyer, Phil Husbands and Inman Harvey 1

How Co-Evolution can Enhance the Adaptive Power of Artificial Evolution:
Implications for Evolutionary Robotics
Stefano Nolfi and Dario Floreano **22**

Running Across the Reality Gap: Octopod Locomotion Evolved in
a Minimal Simulation
Nick Jakobi **39**

Detour Behaviour in Evolving Robots:
Are Internal Representations Necessary?
Orazio Miglino, Daniele Denaro, Guido Tascini and Domenico Parisi **59**

Evolving Robot Behaviours with Diffusing Gas Networks
Phil Husbands **71**

Explaining the Evolved: Homunculi, Modules, and Internal Representation
Michael Wheeler **87**

Some Problems (and a Few Solutions) for Open-Ended
Evolutionary Robotics
Nick Jakobi and Matthew Quinn **108**

Noise and the Pursuit of Complexity: A Study in Evolutionary Robotics
Anil K. Seth **123**

Harware Solutions for Evolutionary Robotics
Dario Floreano and Francesco Mondada **137**

Blurred Vision: Simulation-Reality Transfer of a Visually Guided Robot
Tom M.C. Smith **152**

Learning to Move a Robot with Random Morphology
Peter Dittrich, Andreas Bürgel and Wolfgang Banzhaf **165**

Learning Behaviors for Environmental Modeling by Genetic Algorithm
Seiji Yamada **179**

Evolving and Breeding Robots
Henrik H. Lund and Orazio Miglino **192**

Off-Line Model-Free and On-Line Model-Based Evolution for
Tracking Navigation Using Evolvable Hardware
Didier Keymeulen, Masaya Iwata, Kenji Konaka, Ryouhei Suzuki,
Yasuo Kuniyoshi and Tetsuya Higuchi **211**

Incremental Evolution of Neural Controllers for Robust
Obstacle-Avoidance in Khepera
Joël Chavas, Christophe Corne, Peter Horvai,
Jérôme Kodjabachian and Jean-Arcady Meyer **227**

Evolutionary Robotics: A Survey of Applications and Problems

Jean-Arcady Meyer[1] and Phil Husbands[2] and Inman Harvey[2]

[1] AnimatLab, Departement de Biologie, Ecole Normale Superieure,France
[2] COGS and CCNR, University of Sussex, Brighton, UK

Abstract. This paper reviews evolutionary approaches to the automatic design of real robots exhibiting a given behavior in a given environment. Such a methodology has been successfully applied to various wheeled and legged robots, and to numerous behaviors including wall-following, obstacle-avoidance, light-seeking, arena cleaning and target seeking. Its potentialities and limitations are discussed in the text and directions for future work are outlined.

1 Introduction

In the last few years, several researchers have attempted to bypass the difficulties of hand-coding the control architectures of mobile robots that have to fulfil given missions in unknown, and possibly changing, environments. Because such difficulties stem from the impossibility of foreseeing each problem the robot will have to solve, and from the lack of basic principles upon which human design might rely, these researchers advocate the so-called evolutionary robotics approach, i.e., an automatic design procedure. According to this approach, a robot's controller, and possibly its overall body plan, is progressively adapted to the specific environment and the specific problems it is confronted with, through an artificial selection process that eliminates ill-behaving individuals in a population while favoring the reproduction of better-adapted competitors.

Such a process calls upon some evolutionary procedure such as a genetic algorithm (Goldberg, 1989), an evolution strategy (Schwefel, 1995), or a genetic programming (Koza, 1992) approach. It involves a population of genotypes (i.e., of information that evolves through successive generations) and a phenotype (i.e., the robot's control architecture, its body plan, and its behavior) that is encoded in any one genotype. A dedicated fitness function is used to assess the behavior of each individual in the population and to direct the selection proper. Dedicated operators such as mutation and cross-over give rise to new genotypes in the population and permit robots of ever-increasing fitness to be generated, until the process converges to some local or global optimum. In the majority of applications, the evolutionary procedure is performed in two stages: fitter phenotypes are first sought through specific robot simulations and are then downloaded in turn on a real robot to check their fitness with respect to real world constraints. However, in some other applications, the evolutionary procedure takes place through evaluations performed directly on the robot and fitnesses are directly assessed through real world interactions. In both cases, software controllers can be evolved. They may be implemented as control programs (in a high level language or in machine code), as a variety of production-rule systems, or as neural networks. Finally, within the so-called evolvable hardware approach (Sanchez and Tomassini, 1996; Higuchi et al., 1997), genotypes code for the configuration of hardware controllers and body plans, and fitnesses are also assessed through real world interactions.

In evolutionary robotics, as in many areas of AI, there is much interplay between engineering and scientific goals and outcomes. Some researchers are primarily interested in making better robots, others in sythesizing control systems, artificial nervous systems, whose mechanisms underpin the generation of interesting adaptive behaviours in an artificial creature. The engineer wants to make the thing work well, the scientist wants to understand how it works, trying to abstract general principles, necessary and sufficient conditions and the like. In much of the work covered in this paper the boundary between these two types of endeavour is often blurred. In our view, evolutionary robotics shows great promise in both areas and it is probably beneficial for the two to remain somewhat entwined. This issue will be returned to towards the end of the paper.

Although numerous aspects of the methodology of evolutionary robotics have been tested in 'simulations' where no particular robot was modelled and there was no question of trying out evolved systems in the real world, such research efforts won't be cited in this review paper, which is centered on real robot applications. The robot the most often used in the applications described herein is Khepera, but it will be shown that other robots, including walking robots, have been used as well. This paper will also provide a discussion of the current potentialities and limitations of evolutionary robotics and will end with suggestions for future work.

2 Real Robot Applications

2.1 Khepera

Khepera (Mondada et al., 1993) is a circular miniature mobile robot with a diameter of 55mm, a height of 30mm, and a weight of 70g that is supported by two wheels and two small Teflon balls. In its basic configuration, it is equipped with eight infra-red proximity sensors — six on the front, two on the back — that may also act as visible-light detectors. The wheels are controlled by two DC motors with incremental encoders that move in both directions. It has an on-board 68000 processor and can also be controlled by an off-board computer via a serial link. Its convenient size, ready availability and the fact that it is straightforward to program, has made it a very popular tool for simple autonomous robotics experiments. As in other areas of new-wave robotics, many evolutionary robotics have been carried out on Kheperas, allowing, at least in principle, replication and comparison of results.

Using the Khepsim simulator, Jakobi et al. (1995) evolved both obstacle-avoidance and light-seeking behaviors in Khepera. The simulation was based on a continuous two-dimensional model of the real world physics and allowed the calculation of the dynamics ofthe robot's sensory inputs in response to its motor signals. Recurrent networks of threshold units that were evolved in simulation evoked qualitatively similar behavior on the real robot, especially when the levels of noise present in the simulation had similar amplitudes to those observed in reality.

To evolve the capacity of moving in the environment while avoiding obstacles, Miglino et al. (Miglino et al., 1995a; Lund and Miglino, 1996) used a two-layer feedforward neural network with no hidden units and a fitness function with three components, which were respectively maximized by speed, by moving in a straight line, and by obstacle avoidance. With the help of a genetic algorithm, the synaptic connections and thresholds of the neural controllers were first evolved through simulation. Then, the corresponding networks were downloaded onto a Khepera and proved to be efficient. A similar two-staged approach has been followed by Salomon (1996), who used a (3,6)-Evolution Strategy with self adaptation of the step size (Back and Schwefel, 1993). Likewise, Naito et al. (1997) used a genetic algorithm to configure how a set of 8 logic elements could be connected to each other and to the sensors and motors of the robot. Within this approach, each controller was downloaded on Khepera and its fitness was directly assessed in the real world. With an alternative, and earlier, approach, Floreano and Mondada (1994), allowed the whole evolutionary process to take place entirely on the robot without human intervention. Two-layer Elman neural networks (Elman, 1990) were used as controllers. This architecture consisted of a single layer of synaptic weights from eight sensor units to two motor units, with recurrent connections within the output layer, the same three component fitness function was used. Using the same neuronal architecture and the same fitness function, and in order to study the interactions between associative learning and evolution, Floreano and Mondada (1996a) let evolve the type of the Hebbian rule that was employed by each synapse in the network. Each synapse was thus genetically described by a set of four properties: whether it was driving or modulatory, whether it was excitatory or inhibitory, its Hebbian rule, and its learning rate. Four Hebbian rules could be used: pure Hebbian, postsynaptic, presynaptic, and covariance (Willshaw and Dayan, 1990). Under such conditions, each decoded neural network changed its own synaptic strength configuration according to its genotypic specifications and without external supervision while Khepera interacted with its environment. Experimental results showed that the efficient controllers that evolved exhibited synapses that were continuously changing in a dynamically stable regime. In other words, knowledge in such networks is not expressed by a final stable state of the synaptic configuration, but by a dynamical equilibrium. There are also indications that such plastic neurocontrollers are more resistant to sensor damage than standard static controllers.

Another study of the interactions between associative learning and evolution is that of Mayley (1996) who evolved simple feedforward neural controllers for wall-following in Khepera. In this work also, besides encoding the network's weights, the genome determined whether each weight was plastic or not i.e., whether it might be changed or not by an Hebbian learning process. Experimental results indicated that, as long as there are costs to be paid for the ability to learn, learning is first selected for and then against as evolution progresses, thus illustrating how a learned trait or behavior may become genetically assimilated.

In Floreano and Mondada (1996b) the evolution of a set of behaviors that allowed a Khepera robot to locate a battery charger and periodically return to it so as to increase its chances of survival has been achieved. In this work, the Khepera robot was equipped with two additional sensors. One ambient light sensor was placed under the robot platform pointing downward, so as to detect a black painted area on the floor that was considered as the place where its battery was recharged. Another simulated sensor was used to provide information about the current energy level of the robot's battery. Thus, the input layer of the neural network consisted of twelve receptors each clamped to one sensor: 8 for IR-emitted light, 2 for lateral ambient light, 1 for floor brightness, and 1 for battery charge. The controller architecture was completed with a hidden layer of 5 units with recurrent connections and an output layer of two units, one for each motor. To evaluate the fitness of each individual, each robot started its life with a fully charged battery that was discharged by a fixed amount at every time step and that was instantaneously recharged if the robot happened to pass over the black area. While a given maximum life time was allotted to each robot, a fully discharged battery entailed instantaneous death. The robot's fitness was accumulated at each step during evaluation and called upon two components: the first one was maximized by speed and the second by obstacle avoidance. Although such a fitness function specified neither the location of the battery station, nor the fact that the robot should reach it, the right behavior evolved because the accumulated fitness of each individual depended both on the performance of the robot and on the length of its life.

In the work of Nolfi (1996b) the parameters of a feedforward neural network with no hidden units were evolved to control a Khepera robot that had to explore its environment, to avoid walls and to remain close to a cylindrical target when it found it. The fitness of each controller was assessed through simulation and depended upon the time spent close to the target. Experimental results showed that the evolved individuals were successful in the real world and that, by intensively using an active perception strategy, they could overcome the problem posed by the fact that the walls and the target were hard to distinguish in most cases. As an extension of this work, and in order to study the interactions of individual learning and evolution, Nolfi and Parisi (1997) added two output units to such feedforward controllers. These units served as auto-teaching units (Nolfi and Parisi, 1993) that set the desired values of the two motor-controlling units when, at the beginning of each individual's test period, a backpropagation algorithm was activated. Because testing could be performed either in an environment with dark walls or in an environment with white walls, backpropagation made it possible for a given individual to learn in which environment it was placed and to accordingly adjust during its lifetime the synaptic weights it inherited from the previous generation. Thus, through successive generations, individuals capable of learning more and more rapidly how to find the target evolved.

Using simulations to evolve simple feedforward neurocontrollers that were later downloaded onto a Khepera robot equipped with a gripper module, Nolfi (Nolfi and Parisi, 1995; Nolfi,1996a,1997a,b,c) evolved the task of keeping clear an arena surrounded by walls, in which small cylindrical trash objects were disposed at random. The best results were obtained when the neural controllers exhibited a so-called emergent modular architecture. Within such architecture, the number of available modules, their internal organization, and the mechanisms that determined their interaction were pre-designed and fixed. However, the way each of these modules was used at each time step depended upon the evolved values of each connection weight and bias within the overall architecture. Such values were directly binary encoded in individual genes. Fitnesses were evaluated by counting the number of objects correctly released outside the arena during a given evaluation time. During evolution, individuals capable of simply picking up targets were slightly favoured. Likewise, experience showed that it was useful to artificially increase the number of times the robot encountered another target while carrying an object, in order to force the evolutionary process to select individuals able to avoid targets when the gripper was already holding something.

Researchers at Dortmund University (Nordin and Banzhaf, 1996; Banzhaf et al., 1997) evolved obstacle-avoidance and object-following behaviors in Khepera with a Genetic Programming (Koza, 1992) variant that manipulates machine code directly. Their system uses linear genomes composed of variable length strings of 32 bit instructions for a SUN-4 computer. Each instruction performs arithmetic or logic operations on a small set of registers and may also include a small integer constant of 13 bits at most. The genetic operators are tailored to manipulate genetic code directly. In particular, crossover occurs between instructions and thus changes the order and number of instructions in offspring programs; mutations are allowed to flip bits within instructions. To evolve obstacle avoidance, a fitness function with a negative and positive part was used. The former was the sum of all proximity sensors; the latter was dependent upon wheel speeds and assessed how straight and fast the robot was moving. For object following, the robot's task was to follow moving objects without colliding with them. The corresponding fitness function used values returned by the 4 sensors facing forward, and rewarded individuals capable of both moving towards objects far away and avoiding too close objects. Encouraging preliminary results have been obtained in experiments where the system is using a memory buffer that stores event vectors representing salient sensory-motor situations encountered in the past.

Instead of directly evolving a complex behavior as a whole, Lee et al. (1997a,b) evolved behavior primitives and behavior arbitrators for a Khepera robot that had to push a box toward a goal position indicated by a light source. To accomplish this task, they used a genetic programming system that evolved the controller programs of two behavior primitives, box pushing (keep pushing a box forward) and box-side-following (move along the side of a box). In addition, they also evolved an arbitrator program that was used to arrange the executing sequence of the behavior primitives. Experimental results show that controllers evolved in simulation were transferred to the real robot without loss of performance.

Several research efforts have aimed at evolving neural controllers for the Khepera robot through developmental approaches that call upon various biomimetic processes — like cell division, cell differentiation, or cell adhesion — to gradually build a neural control architecture. Controllers for obstacle-avoidance, light-seeking or light-avoiding behaviors have thus been evolved by Eggenberger (1996). Wall-following and obstacle-avoidance behaviors have also been evolved through such a developmental approach by Michel (Michel, 1996; Michel and Collard,1996).

Smith (1997) successfully evolved a football playing Khepera. The Khepera was equipped with a minimal 1-D CCD camera-based visual system and used this to guide its behaviour. Behaviours evolved in simulation allowed the robot to successfully find the ball and accurately push it up the pitch and into the goal. When down-loaded onto the real Khepera, the controllers were equally successful. A GA was used to set the weights on a fixed architecture neural network in which 16 visual inputs, recurrent connections from the two motor outputs and the input from a crude compass, were all fully connected to 16 hidden units. The hidden units received input from a bias unit and each had recurrent connections. Each of the 16 hidden units was connected to both left and right motor neurons.

Finally, with the aim of evolving a behavior that was at least one step up from the simple reactive behaviors that have been sought so far, Jakobi (1997a,b) succeeded in evolving reliably fit recurrent neural network controllers that allowed a Khepera robot to memorize on which side of a corridor it passed through a beam of light. Then, when the robot arrived at a T-maze junction at the end of the corridor, its task was to turn in the direction of the memorized light and move down the corresponding arm. Controllers that have been evolved within around 1000 generations in simulation were downloaded onto Khepera and performed the task satisfactorily and efficiently. Both this and Smith's footballing Khepera were evolved using ultra-fast ultra-minimal simulations (Jakobi 1998).

2.2 Other robots

Several experiments have been performed at Sussex University (Harvey et al., 1994; Husbands et al., 1997, Jakobi, 1997a,b) in which discrete-time dynamical recurrent neural networks and visual sampling morphologies are concurrently evolved to allow a gantry robot to perform various visually guided tasks. Such experiments called upon a CCD camera sensing its environment through a swiveling mirror. For instance, within an environment predominantly dark, the robot had to move toward fixed or mobile white targets. Likewise, in one experiment it had to approach a white triangle while ignoring a white rectangle. In such experiments, successful behaviors were evolved using a genetic algorithm acting on pairs of chromosomes encoding the visual morphology and the neural controller of the robot. One of the chromosomes was a fixed length bit string encoding the position and size of three visual receptive fields from which the visual signals processed by the neural controller were calculated. The other was a variable length character string encoding the number of hidden units and the number of excitatory and inhibitory connections between neurons. The number of input nodes was fixed to seven (one input for each of three visual receptive field and for each of four tactile sensors) and the number of output nodes was fixed to four (two for each 'virtual wheel', whose motions were translated into gantry movements and mirror angular velocities); the hidden nodes were variable in number. Unlike most of the work previously mentioned, in this research the network architecture was not constrained; arbitrarily recurrent networks of any topology were allowed. The methodology followed was also rather different from that practiced elsewhere; a converged population was taken through an incrementally more difficult succession of environments using different fitness functions at each stage (Harvey 1992, Husbands and Harvey, 1992). The apparatus was designed to allow real-world evolution (Harvey et al., 1994) but behaviours have also been successfully evolved in minimal simulations (Jakobi, 1997a). In this later work the number of visual inputs was not fixed and the lighting conditions were far noisier than in the original experiments. Highly robust target discriminator controllers were evolved.

Interactions between reinforcement learning and evolution have been exploited in the work by Grefenstette and Schultz (1994), which calls upon the use of the SAMUEL classifier system (Grefenstette and Cobb, 1991) for evolving collision-free navigation in a Nomad 200 mobile robot equipped with 20 tactile, 16 sonar, and 16 infra-red sensors. Within such an approach, apart from being liable to mutation, the condition part of each of SAMUEL's rule which was compared against the current sensor readings was also submitted to dedicated generalization and specialization operators. The task consisted of learning to reach a fixed goal location in a predetermined time, starting from a fixed initial position within an environment that contained obstacles whose positions were changed at each trial. With a population size of 50 rules, rule sets evaluated through simulation over 50 generations were downloaded on the robot and proved to be efficient 86similar approach is that of Colombetti and Dorigo (1993) who used the ALECSYS software tool (Dorigo, 1993) to evolve the control architecture of the AutonoMouse, a mouse-shaped autonomous robot equipped with two on/off eyes positioned in front of the robot and sensing light within a cone of about 60 degrees. In this work, the robot's control architecture was a set of interconnected classifier systems and the behavior to evolve was light-chasing. To succeed, the robot had to learn appropriate moves so as to cope with situations where the target light was on, but out of the robot's sight. The robot's fitness was evaluated through light intensity, detected by a dedicated central light sensor.

Miglino et al. (1995b) evolved a four-layer Elman-like recurrent neural networks with 2 sensory units, 2 output units, 2 hidden units, and 1 memory unit that allowed a mobile Lego robot to explore the greatest percentage of an open area within an allotted number of steps. Two optosensors were used to detect whether the areas ahead and behind the robot's current location were black or white, thus allowing the robot to move within a central white surface surrounded by a black border. Such moves were determined by the values of the two output units. The architecture of the controllers was fixed and only the weights of the connections were encoded in the genotype, as a vector of 17 integer numbers. Although the fitness of each controller was assessed through simulations, experiments showed that evolved controllers were efficient in the real world, despite the fact that the real trajectories were significantly different from the simulated ones.

Yamauchi and Beer (1994) used a Nomad 200 mobile robot to evolve neural controllers capable of identifying one of two landmarks based on the time-varying sonar signals received as the robot turned around the landmark. The robot's trajectory was controlled by a fixed behavior-based control system that allowed the robot to find a wall and follow it counterclockwise around the perimeter of the experimental room. A single sonar on the left side of the robot was used to detect a central landmark and its range signals were input to each of the eight neurons in a continuous-time fully-connected recurrent neural network (Beer and Gallagher, 1992). One of these neurons was designated the output unit and its firing rate after a fixed period of time (i.e., after the input signal sequence has been integrated over time) was used to classify the landmark. Network parameters — like time constants, thresholds, or connection weights — were genetically encoded as vectors of real numbers, of which each element was indivisible under crossover. The fitness function of each individual in a population of 100 networks was evaluated in simulation and assessed the average capacity of the network to correctly identify the landmarks over six test trials. After 15 generations, an individual capable of correctly recognizing the landmarks in simulation was generated. When transferred onto the real robot, it correctly classified the landmarks in 17 out of 20 test trials.

In Yamauchi (1993), other evolutionary robotic simulations are described that have been successfully applied to predator avoidance in a Nomad 200 robot. In this approach, dynamic neural networks were used to perform the task of evading a moving pursuer while avoiding collisions with stationary obstacles.

Baluja (1996) presents an evolutionary method for designing a neural controller for the Carnegie Mellon's NAVLAB autonomous land vehicle. To assess its steering abilities, the neural network is shown video images from the NAVLAB's onboard camera as a person drives and its task is to output the direction in which the person is currently steering. A maximal network architecture is defined, which determines the structure and maximum connectivity of the controller to which, during evolution, connections may be removed but not added. In one series of experiments, this maximal network architecture was a fully-connected perceptron with a 15 x 16 pixels input retina, a five unit hidden layer, and a single unit layer whose activation determined how sharply the steering should be to the left or to the right of center. In a second series of experiments, the same architecture was used, but with 30 output units, each of which was considered as representing the network's vote for a particular steering direction. In both cases, the so-called PBIL (Population-Based Incremental Learning) evolutionary algorithm was used, according to which a probability vector is evolved as a prototype from which potentially highly fit networks can be derived. This vector specifies the probabilities of having a 1 or a 0 in each bit position of a string encoding the topology and connection weights of a neural controller. During evolution, in a manner similar to the training of a competitive learning network, the values in the probability vector are progressively shifted toward the bit values that specify efficient network designs. This evolutionary approach performed better, on average, than standard backpropagation, especially in the one-output networks.

Using a genetic algorithm acting on individuals represented as real-coded vectors of weights, Meeden (1996) evolved recurrent neural controllers for a four-wheeled robot that had to continually keep moving, to avoid contacts with walls, and either to seek or avoid light depending upon its current goal. This robot was equipped with three front and one back touch sensors, with two light sensors, and with one goal sensor that indicated that the robot should seek out (or avoid) the light until a maximum (or minimum) light reading was obtained. For movement, the robot had two servo-motors: one controlling forward and backward motion, the other controlling steering. Elman-like networks with a fixed architecture were used for that purpose — with 7 input units each connected to a given sensor, 5 hidden units with recurrent self-connections, and 4 output units that determined how to set the motors for the next time step. During evaluation, the fitness of a given controller was incremented or decremented after each robot's action, according to a reward scale that took into account whether or not the robot accomplished a light goal, kept moving, had any touch sensor triggered, and correctly followed the light gradient. Experimental results showed that the evolutionary update of weights out-performed a complementary reinforcement backpropagation learning algorithm (Ackley and Litman, 1990) under delayed reinforcement conditions, i.e., when no light gradient reinforcement was provided between two switching-goal episodes.

Jeong and Lee (1997) got promising results suggesting that a genetic algorithm could be used to automatically design the controllers and the control strategies for two-wheeled soccer playing robots. Such robots are assumed to be used within an experimental setup consisting of a host computer that processes the vision data acquired by a camera and sends to each robot information about the positions of the ball and of each robot. A two-stage evolutionary approach has been investigated. In a first stage, production rules have been evolved, whose condition parts take into account the positions of the relevant objects i.e., the partners, the opponents, the goals, and the ball and whose action parts trigger a relevant action i.e., a move, a dribble or a kick. In a second stage, optimal on-off control signals to the motors were evolved that allowed a robot to reach a position with desired coordinates and orientation.

Ram et al. (1994) used a genetic algorithm to find appropriate combinations of parameters for basic reactive behaviour schemas used to control an autonomous mobile robot engaged in navigation tasks. Example primitive behaviours are: move-to-goal and avoid-static-obstacle. Parameters involved in the underlying implementation of these behaviours are quantities such as: goal gain (strength with which robot moves towards goal), obstacle gain (strength with which robot moves away from obstacle) and obstacle sphere of influence (distance from obstacle at which robot is repelled). The use of a genetic algorithm greatly reduced the time required to configure the navigation systems.

Gallagher et al. (1996) describe experiments were neural networks controlling locomotion in an artificial insect were evolved in simulation and then successfully downloaded on a real 6-legged robot. In this approach, each leg was controlled by a fully interconnected network of 5 Hopfield-like continuous neurons (Hopfield, 1984), each receiving a weighted sensory input from that leg's angle sensor. Three of these neurons were motor neurons that respectively governed the state of the forward and backward joint torques of the leg and the state of the corresponding foot. The remaining two neurons were interneurons with no pre-specified role. Thanks to various simplifying assumptions (Beer and Gallagher, 1992), a set of only 50 parameters which described neuronal physical constants, crossbody connection weights and intersegmental connection weights needed to be encoded in the insect's genotype as mere bit strings.

A genetic algorithm has been used by Galt et al. (1997) to derive the optimal gait parameters for a Robug III robot an 8-legged, pneumatically powered walking and climbing robot. The individual genotypes were encoded to represent the phase and duty factors i.e., the coordinating parameters that represent each leg's support period and the time relationships between the legs. Controllers were thus evolved that have been proved capable of deriving walking gaits that are suitably adapted to a wide range of terrains, damages or system failures. Future research will be targeted at using information on the terrain contours provided by the robot's legs. Such information can be used by neural networks to provide one step ahead forecast of the terrain conditions and hence improve the walking efficiency.

Gomi and Ide (1997) evolved the gaits of an 8-legged OCT-1 robot (AAI Systems, Inc.) by loading it with a set of 50 software invoked control processes that are each given in turn a fixed amount of time to actuate the robot's legs. The corresponding genotypes are made of 8 similarly organized sets of genes, each gene coding for a legs motion characteristics such as the amount of delay after which the leg begins to move, the direction of the leg's motion, the end positions of both vertical and horizontal swings of the leg, the vertical and horizontal angular speed of the leg, etc. The fitness function is set in favor of a robot that stands up, evolves coordination among its legs motions, and has a tendency to move forward. Moreover, fitness scores are increased when internal sensors monitoring the servo motor electric currents indicate that a given leg is moved under proper loading conditions. Fitness scores are decreased when any of the sensors located on the belly of the robot detects a contact with the floor. Typically, after generation 10, most individuals succeed in standing and walking with a faint gait. Likewise, after a few dozen generations, a mixture of tetrapod and wave gaits is obtained.

Gruau and Quatramaran (1997) also evolved controllers for walking in the OCT-1 robot. Using a developmental approach called Cellular Encoding (Gruau, 1995) i.e., an approach that genetically encodes a grammar-tree program that controls the division of cells growing into a discrete-time dynamical recurrent neural network they first evolved a single-leg neural controller with one input and two outputs. When commands for return stroke or power stroke were input to the controller, it succeeded in respectively lifting the foot up and propelling the leg forward or puting the foot down and propelling the leg backwards. Then, they put together 8 copies of the leg controller and evolved a neural network that called upon 8 oscillators with correct frequency, coupling, and synchronization, which generated a smooth and fast quadripod locomotion gait. Gruau used a form of interactive evolution, in which the experimenter decides fitness scores by observing candidate robot behaviours, and shapes the course of evolution by favouring certain traits they feel will be useful in the final solution. Jakobi (1998) has successfully used his minimal simulation techniques to evolve controllers for the same 8-legged robot. He used networks very similar to those employed by Beer and Gallagher (1992). Evolution in simulation took about 2 hours only, and then transferred perfectly to the real robot. His neural controllers allowed the robot to avoid obstacle using a fluid combination of forward and backward gaits.

3 Evolvable Hardware

Evolved Hardware controllers are not programmed to follow a sequence of instructions, they are configured and then allowed to behave in real time according to semiconductor physics.

Thompson (1995,1997) used artificial evolution to design hardware circuits as an on-board controllers for two-wheeled autonomous mobile robots displaying simple wall-avoidance behavior in an empty arena. This work is now possible using particular types of Field Programmable Gate Arrays (FPGAs) which are appropriate for evolutionary applications (eg. the Xilinx XC6200 series). A FPGA is a Very Large Scale Integration (VLSI) silicon chip containing a large array of components and wires. Switches distributed throughout the chip can be set by an evolutionary algorithm and determine how each component behaves and how it connects to the wires. In the 1995 work before the appropriate FPGAs were available, it was necessary to construct ones own equivalent reconfigurable circuits. Thompson's approach called upon a so-called DSM (Dynamic State Machine) equipped with genetic synchronizers and with a global clock whose frequency was also under genetic control. Thus evolution determined whether each signal was passed straight through asynchronously, or whether it was synchronized according to the global clock. This process took place within the robot in a kind of "virtual reality" in the sense that the real evolving hardware controlled the real motors, but the wheels were just spinning in the air. The movement that the robot would have actually performed if the wheels had been actually supporting it were then simulated and the sonar echo signals that the robot was expected to receive were supplied in real time to the hardware DSM. Excellent performances were attained after 35 generations, with good transfer from the virtual environment to the real world. In later work after the Xilinx XC6200 chips became first available, similar results (using infra-red rather than sonar) have been obtained with a Khepera robot equipped with an onboard Field-Programmable Gate Array (FPGA) (Thompson, 1997).

Using a Boolean function approach implemented on gate-level evolvable hardware, Keymeulen et al. (1997a,b) evolved a navigation system for a mobile robot capable of reaching a colored ball while avoiding obstacles during its motion. The mobile robot was equipped with infra-red sensors and an active vision system furnishing the direction and the distance to the colored target. A programmable logic device (PLD) was used to implement a Boolean function in its disjunctive form, which has been proved to be sufficient to control tracking-avoiding tasks (Lund and Hallam, 1997). It appeared that such gate level evolvable hardware was able to take advantage of the correlations in the input states and to exhibit useful generalization abilities, thus allowing the simulated evolution of a robust behavior in simple environments and a good transfer into the real world. Future work aims at accelerating on-line evolution by allowing the robot to do some experimentation in an internal model of its environment, to be implemented in an additional special purpose evolvable system.

Finally, Lund et al. (1997) advocate the use of so-called true evolvable hardware to evolve, not only a robot's control circuit, but also its body plan, which might include the types, numbers and positions of the sensors, the body size, the wheel radius, the motor time constants, etc. These authors are currently developing a new piece of reconfigurable hardware that will make it possible to co-evolve the control mechanisms and the auditory morphology of a Khepera robot behaving like a female cricket which is able to use phonotaxis to locate a song emitting male.

4 Discussion

The research field of evolutionary robotics came into being in the early 1990s (e.g. Husbands and Harvey, 1992; Brooks, 1992), and has expanded rapidly, largely in Europe and Japan. The special requirements of an evolutionary approach, in particular large numbers of trials, raise particular problems. These were first tackled with a purpose-built piece of hardware at Sussex (Harvey et al, 1994), but then the field really took off with the introduction of the Khepera robot built in Lausanne (Mondada et al., 1993). This allowed many research laboratories to move into the field relatively cheaply, and generate results that are replicable elsewhere by others with the same class of robot.

An evolutionary approach to design could in principle be applied to any class of control system architectures, but it is significant that the great majority of research reported here uses some form of neural network. Classifier systems have also been used, but these can usually be reconceptualised as implementing something functionally equivalent to a neural network (Miller and Forrest, 1989). Likewise at least some of the evolvable hardware approaches (eg. Thompson 1997) treat the hardware as electrical circuits (loosely comparable to neural circuits) rather than as implementing Boolean functions or computational rules. So for the most part the research reported here has moved away from world-modelling classical AI ideas on robotics (Moravec, 1983). Where neural networks are recurrent and incorporate some reference to time scales, then these fit in with the Dynamical Systems approach to cognition (Beer, 1995; Van Gelder, 1995).

As mentioned earlier, evolutionary robotics has potential in research with scientific aims as well as that with more exclusively engineering goals. Two of the ways evolutionary robotics can be used in cognitive science are: as a means to explore spaces of behaviour generating mechanisms and architectures; and as a way of synthesizing adaptive artificial nervous systems using as few preconceptions as possible about how a given behaviour should be generated. On analysis, evolved controllers may well make use of very different kinds of mechanisms from those postulated by conventional cognitive science. See (e.g. Wheeler, 1996; Hendriks-Jansen, 1996; Beer, 1996; Cliff and Nobel, 1997) for further discussion of issues surrounding this topic. A similar role is starting to be explored in neuroscience; spaces of postulated mechanisms can be searched for plausible candidates. Such spaces might usefully range from possible high-level behavioural strategies to the potential roles of secondary messengers in neuronal networks.

This paper has concentrated exclusively on research that has been implemented on real robots. There have also been many studies in the artificial life and animat literature of artificial agents constructed to 'live' in artificial virtual worlds with varying degrees of resemblance to the real world. There is probably a consensus amongst those who have worked both with real robots and with simulations of robots that most of the really hard problems in robotics cannot be appreciated by those who have worked solely with simulations. This raised question marks as to whether simulations were of any use at all. Some of these worries have been resolved by the several pieces of research reported here where control systems evolved partly or wholly within careful simulations did indeed behave appropriately when downloaded onto the real robot. There is a study of the necessary relationships between simulations and reality in (Jakobi 1998). Adequate simulations, particularly ones that are significantly faster and cheaper than testing on a real robot, are potentially significant if evolutionary robotics is to be economic for practical applications. The scientific work mentioned in the previous paragraph will often be most convincing when real robots, facing real noisy and uncertain worlds, are used. However, aspects of it may well make use of abstract computer models when assumptions and simplifications are carefully and appropriately drawn.

Irrespective of the aims and style of individual pieces of evolutionary robotics research, no such endeavour will progress significantly unless a number of key *interacting* problems are addressed. These include:

- What is the most appropriate type of genotype to phenotype mapping to use for a given class of desired behaviours?

- What kinds of basic nervous system building blocks should be used?
- How is fitness best evaluated?
- What kind of evolutionary algorithm is best suited to a particular evolutionary robotics project?

These items are briefly discussed below. It would be premature to say anything very concrete at the moment. However, various researchers are exploring the interwoven strands of these central problems. As knowledge and experience is built up and exchanged, it is hoped that the best way forward will become clear. Currently there is no lack of ideas on possible directions, which makes for a healthy and interesting research field.

A wide variety of evolutionary algorithms have been used in the research covered here. There is as yet no clear consensus on which type of evolutionary algorithm, and which parameter settings, are appropriate for particular problems, and much experimentation is on a trial and error basis. A significant division in evolutionary robotics lies between those approaches where the form of the control structure (e.g. layers and nodes of a neural network) are predefined by the researcher, leaving such variables as connection weights to be determined through evolution; and those where the very structure and form of the control system is to be evolved. In the former class, evolutionary algorithms are just one possibility amongst many feasible optimisation techniques, and are typically working with a fixed number of real-valued variables representing connection weights; these can be encoded directly on the genotype as real numbers, or encoded in binary or other form. For the second class perhaps evolutionary algorithms come into their own, as there are few alternative techniques for open-ended search through a space of possible structures. Here the role of the genotype-to-phenotype mapping is of great significance, as different methods of morphogenesis may have differing suitability for evolution. It is likely that the control networks and sensor morphologies needed for more complex behaviours will require various forms of large scale structure, including repeated sub-elements. It is difficult to see how this could be achieved without recourse to a fairly sophisticated developmental genotype to phenotype mapping. See (Kodjabachian and Meyer, 1994; Husbands et al., 1997) for discussions of encoding issues.

Although we a very long way from an understanding of real brains, it can at least be said that both vertebrate and invertebrate nervous systems are complex highly heterogeneous dynamical systems with a number of distinct but interacting processes at play. These include electrical, short-range chemical and long-range diffusing chemical mechanisms. Diffusing gases can modulate the intrinsic properties of nerve cells and synapses, sometimes causing radical changes (Garthwaite, 1991; Salter et al., 1991). It appears that these effects can be short-lived or permanent (through changes in cells at the genetic level). This is all a very far cry from the connectionist style networks favoured by most evolutionary roboticists. Why are natural nervous systems so complex (or at least appear to be so complex)? Could it be that systems capable of generating sophisticated adaptive behaviour require the kinds of intricate and deeply entwined mechanisms that we observe in nature? Could it be that this class of dynamical systems is more evolvable? Whatever the answer, it is clear that there is a huge space of possible network types to explore, with varying degrees of plasticity and dynamical complexity. It certainly seems a mistake to imagine that simple connectionist style networks can take us very far, although exactly how far they can take us is itself an interesting question.

Evaluating robot behaviours brings its own special problems. Many behaviours are inherently noisy which must be taken into account in designing fitness criteria. Every effort must be made to eliminate the possibility of giving too much credit to 'lucky' controllers that perform very well in some circumstances but poorly under most conditions. Very often multiple trials are used for each fitness evaluation, so that a representative spread of conditions is encountered by each robot. As attempts are made to evolve more complex behaviours the issue of how to design fitness criteria will become more pertinent. It is possible that rather implicit criteria that map 'good behaviour' onto 'survival' or 'maintaining viability' will be necessary.

For evolutionary robotics to have some practical applications, the time and expense of multiple evaluations of robot control systems must be minimised. Apart from the use of simulations where viable, elements of *all* the above (encoding, network type, evaluation criteria, evolutionary algorithm) will be important. In order to make advances in understanding how to evolve more complex behaviours faster, it will be necessary to understand more about the dynamics of the evolutionary process itself, the properties of encoding schemes, the behaviour generating power of particular types of networks, and how all these combine to produce search spaces that are more or less amenable to evolution.

5 Conclusion

As an infant research field, evolutionary robotics is a relatively thriving baby with much research going on in parallel across many research groups. Some basic achievements have been reached with real robots, typically on fairly simple robot behaviours which are often comparable to those achieved by more orthodox methods. Enough experience has been built up for the start of a clear understanding of the relationships between simulations and reality. The field is starting to move from reports of one-off successes towards repeatable results. The challenge is to move from basic robot behaviours to ever more complex, non-reactive ones. There is much to be done, go to it.

References

Ackley, D.H. and Littman, M.L. 1990. Generalization and scaling in reinforcement learning. In Touretzky (Ed.). Advances in Neural Information Processing Systems 2. Morgan Kaufmann.

Back, T. and Schwefel, H.P. 1993. An Overview of Evolutionary Algorithms for Parameter Optimization. Evolutionary Computation, 1,1,1–23.

Baluja, S. 1996. Evolution of an Artificial Neural Network Based Autonomous Land Vehicle Controller. IEEE Transactions on Systems, Man, and Cybernetics Part B: Cybernetics. 26, 3, 450–463.

Banzhaf, W., Nordin, P. and Olmer, M. 1997. Generating Adaptive Behavior for a Real Robot using Function Regression within Genetic Programming. In Koza et al. (Eds.). Genetic Programming 1997. Morgan Kaufmann.

Beer, R.D. and Gallagher, J.C. 1992. Evolving Dynamical Neural Networks for Adaptive Behavior; Adaptive Behavior, 1, 1, 91–122.

Beer, R. 1995. A dynamical systems perspective on agent-environment interaction. Artificial Intelligence, 72, 173–215.

Beer, R. 1996. Toward the evolution of dynamical neural networks for minimally cognitive behaviour. In: from Animals to Animats 4, P. Maes et al. (Eds.), MIT Press/Bradford Books, 421–429.

Brooks, R. 1992. Artificial Life and Real Robots. In: Proceeding of First European Conference on Artificial Life, F. Varela and P. Bourgine (Eds.), 3–10, MIT Press/Bradford Books.

Cliff, D., Harvey, I. and Husbands, P. 1993. Explorations in evolutionary Robotics. Adaptive Behavior. 2,1, 73–110.

Cliff, D. and Noble, J. 1997. Knowledge-based vision and simple visual machines. Philosophical Transactions of the Royal Society: Biological Sciences, 352(1358), 1165–1175.

Colombetti, M. and Dorigo, M. 1993. Learning to Control An Autonomous Robot By Distributed Genetic Algorithms. In Meyer, Roitblat and Wilson (Eds.). Proceedings of the Second International Conference on Simulation of Adaptive behavior: From Animals to Animats 2. The MIT Press/Bradford Book.

Dorigo, M. 1993. Genetic and non-genetic operators in ALECSYS. Evolutionary Computation. 1(2), 151–164.

Eggenberger, P. 1996. Cell Interactions as a Control Tool of Developmental Processes for Evolutionary Robotics. In Maes, Mataric, Meyer, Pollack and Wilson (Eds.). Proceedings of the Fourth International Conference on Simulation of Adaptive behavior:From Animals to Animats 4. The MIT Press/Bradford Book.

Elman, J.L. 1990. Finding structure in time. Cognitive Science. 2, 179–211.

Floreano, D. and Mondada, F. 1994. Automatic Creation of an Autonomous Agent: Genetic Evolution of a Neural-Network Driven Robot. In Cliff, Husbands, Meyer and Wilson (Eds.). Proceedings of the Third International Conference on Simulation of Adaptive behavior: From Animals to Animats 3. The MIT Press/Bradford Book.

Floreano, D. and Mondada, F. 1996a. Evolution of plastic neurocontrollers for situated agents. In Maes, Mataric, Meyer, Pollack and Wilson (Eds.). Proceedings of the Fourth International Conference on Simulation of Adaptive behavior: From Animals to Animats 4. The MIT Press/Bradford Book.

Floreano, D. and Mondada, F. 1996b. Evolution of Homing Navigation in a Real Mobile Robot. IEEE Transactions on Systems, Man, and Cybernetics Part B: Cybernetics. 26, 3, 396–407.

Gallagher, J.C., Beer, R.D., Espenschield, K.S. and Quinn, R.D. 1996. Application of evolved locomotion controllers to a hexapod robot. Robotics and Autonomous Systems. 19, 95–103.

Galt, S., Luk, B.L. and Collie, A.A. 1997. Evolution of Smooth and Efficient Walking Motions for an 8-Legged Robot. Proceedings of the 6th European Workshop on Learning Robots. Brighton, UK.

Garthwaite, J. 1991. Glutamate, nitric oxide and cell-cell signalling in the nervous system. Trends in Neuroscience, 14, 60–67.

Goldberg, D. E. 1989. Genetic Algorithms in Search, Optimization and Machine Learning. Addison-Wesley.

Gomi, T. and Griffith, A. 1996. Evolutionary Robotics An Overview. Proceedings of the IEEE 3rd International Conference on Evolutionary Computation. IEEE Society Press.

Gomi, T. and Ide, K. 1997. Emergence of gaits of a legged Robot by Collaboration through Evolution. Proceedings of the International Symposium on Artificial Life and Robotics. Springer Verlag.

Grefenstette, J. and Cobb, H.C. 1991. User's guide for SAMUEL (NRL Memorandum Report 6820). Washington, DC: Naval Research Laboratory.

Grefenstette, J. and Schultz, A. 1994. An evolutionary approach to learning in robots. Proceedings of the Machine Learning Workshop on Robot Learning. New Brunswick, NJ.

Gruau, F. 1995. Automatic definition of modular neural networks. Adaptive Behavior. 3, 2, 151–183.

Gruau, F. and Quatramaran, K. 1997. Cellular Encoding for Interactive Evolutionary Robotics. In Husbands and Harvey (Eds.). Fourth European Conference on Artificial Life. The MIT Press/Bradford Books.

Harvey, I. 1992. Species Adaptation Genetic Algorithms: The Basis for a Continuing SAGA. In Proceedings of the First European Conference on Artificial Life, F. Varela and P. Bourgine (Eds.), 346-354, MIT Press/Bradford Books, Cambridge, MA.

Harvey, I., Husbands, P. and Cliff, D. 1994. Seeing The Light: Artificial Evolution, Real Vision. In Cliff, Husbands, Meyer and Wilson (Eds.). Proceedings of the Third International Conference on Simulation of Adaptive behavior: From Animals to Animats 3. The MIT Press/Bradford Book.

Harvey, I., Husbands, P., Cliff, D., Thompson, A. and Jakobi, N. 1997. Evolutionary Robotics: The Sussex Approach. Robotics and Autonomous Systems. 20, 205–224.

Higuchi, T., Iwata, M. and Liu, W. (Eds.). 1997. Evolvable Systems: From Biology to Hardware. Springer Verlag.

Hopfield, J.J. 1984. Neurons with graded response properties have collective computational properties like those of two-state neurons. Proceedings ofthe National Academy of Sciences, USA, 81, 3088–3092.

Hendriks-Jansen, H. 1996. Catching Ourselves in the Act: Situated Activity, Interactive Emergence, Evolution, and Human Thought. MIT Press/Bradford Books.

Husbands, P. and Harvey, I. 1992. Evolution versus Design: Controlling Autonomous Robots. In: Integrating Perception, Planning and Action, Proceedings of 3rd Annual Conference on Artificial Intelligence, Simulation and Planning, 139–146, IEEE Press.

Husbands, P, Harvey, I., Cliff, D. and Miller, G. 1994. The Use of Genetic Algorithms for the Development of Sensorimotor Control Systems. In Nicoud and Gaussier (Eds.).From Perception to Action. IEEE Computer Society Press.

Husbands, P, Harvey, I., Cliff, D. and Miller, G. 1997. Artificial Evolution: A New Path for AI? Brain and Cognition, 34, 130–159.

Jakobi, N. 1998. Minimal Simulations for Evolutionary Robotics. D. Phil. Thesis, School of Cognitive and Computing Sciences, University of Sussex.

Jakobi, N. 1997a. Half-baked, Ad-hoc and Noisy: minimal Simulations for Evolutionary Robotics. In Husbands and Harvey (Eds.). Fourth European Conference on Artificial Life. The MIT Press/Bradford Books.

Jakobi, N. 1997b. Evolutionary Robotics and the Radical Envelope of Noise Hypothesis. Adaptive Behavior. 6,1, 131–174.

Jakobi, N., Husbands, P. and Harvey, I. 1995. Noise and the reality gap: The use of simulation in evolutionary robotics. In Moran, Moreno, Merelo and Chacon (Eds.). Advances in Artificial Life: Proceedings of the Third European Conference on Artificial Life. Springer Verlag.

Jeong, I.K. and Lee, J.J. 1997. Evolving cooperative mobile robots using a modified genetic algorithm. Robotics and Autonomous Systems,21, 197–205.

Kodjabachian, K. and Meyer, J-A. 1994. Development, Learning and Evolution in Animats. In: Proceedings of From Perception to Action Conference, P. Gaussier and J-D. Nicoud (Eds.), IEEE Computer Society Press, 96–109.

Keymeulen, D., Durantez, M., Konaka, M., Kuniyoshi, Y. and Higuchi, T. 1997a. An Evolutionary Robot Navigation System Using a Gate-Level Evolvable Hardware. In Higuchi, Iwata and Liu (Eds.). Evolvable Systems: From Biology to Hardware. Springer.

Keymeulen, D., Konaka, K., Iwata, M., Kuniyoshi, Y. and Higuchi, T. 1997b. Robot Learning using gate-Level evolvable hardware. Proceedings of the 6th European Workshop on Learning Robots. Brighton, UK.

Koza, J. 1992. Genetic Programming. The MIT Press.

Lee, W.P., Hallam, J. and Lund, H.H. 1997a. Applying Genetic Programming to Evolve Behavior Primitives and Arbitrators for Mobile Robots. Proceedings of IEEE Fourth International Conference on Evolutionary Computation. Piscataway, NJ.

Lee, W.P., Hallam, J. and Lund, H.H. 1997b. Learning Complex Robot Behaviours by Evolutionary Approaches. Proceedings of the 6th European Workshop on Learning Robots. Brighton, UK.

Lund, H.H. and Hallam, J. 1997. Evolving sufficient robot controllers. Proceedings of IEEE Fourth International Conference on Evolutionary Computation. Piscataway, NJ.

Lund, H.H., Hallam, J. and Lee, W.P. 1997. Evolving Robot Morphology. Proceedings of IEEE Fourth International Conference on Evolutionary Computation. Piscataway, NJ.

Lund, H.H. and Miglino, O. 1996. From Simulated to Real Robots. Proceedings of the 3rd IEEE International Conference on Evolutionary Computation. IEEE Computer Society Press.

Mataric, M. and Cliff, D. 1996. Challenges in evolving controllers for physical robots. Robotics and Autonomous Systems. 19, 67–83.

Mayley, G. 1996. The Evolutionary Cost of Learning. In Maes, Mataric, Meyer, Pollack and Wilson (Eds.). Proceedings of the Fourth International Conference on Simulation of Adaptive behavior: From Animals to Animats 4. The MIT Press/Bradford Book.

Meeden, L.A. 1996. An Incremental Approach to Developing Intelligent Neural Network Controllers for Robots. IEEE Transactions on Systems, Man, and Cybernetics Part B: Cybernetics. 26, 3, 474–485.

Michel, O. 1996. An Artificial life Approach for the synthesis of Autonomous Agents. In Alliot, Lutton, Ronald, Schoenauer and Snyers (Eds.). Artificial Evolution. Springer.

Michel, O. and Collard, P. 1996. Artificial Neurogenesis: An application to Autonomous Robotics. In Radle (Ed.). Proceedings of The 8th. International Conference on Tools in Artificial Intelligence. IEEE Computer Society Press.

Miglino, O., Lund, H.H. and Nolfi, S. 1995a. Evolving Mobile Robots in Simulated and Real Environments. Artificial Life, 2, 417–434.

Miglino, O., Nafasi, K. and Taylor, C. 1995b. Selection for Wandering Behavior in a Small Robot. Artificial Life, 2, 101–116.

Miller, J. and Forrest, S. 1989. The Dynamical Behavior of Classifier Systems. In: Proceedings of the Third International Conference on Genetic Algorithms, J. Schaffer (Ed.), Morgan Kaufmann, 304–310.

Mondada, F., Franzi, E. and Ienne, P. 1993. Mobile robot miniaturization: A tool for investigation in control algorithms. Proceedings of the Third International Symposium on Experimental Robotics. Kyoto, Japan.

Moravec, H. 1983. The Stanford Cart and the CMU Rover. Proc. of IEEE", vol. 71, 872–884.

Naito, T., Odagiri, R., Matsunaga, Y., Tanifuji, M. and Murase, K. 1997. Genetic Evolution of a Logic Circuit Which Controls an Autonomous Mobile Robot. In Higuchi,Iwata and Liu (Eds.). Evolvable Systems: From Biology to Hardware. Springer.

Nolfi, S. 1996a. Evolving non-Trivial Behaviors on Real Robots: a garbage collecting robot. Technical Report, Institute of Psychology, CNR,Rome.

Nolfi, S. 1996b. Adaptation as a more powerful tool than decomposition and integration. Technical Report, Institute of Psychology, CNR, Rome.

Nolfi, S. 1997a. Using Emergent Modularity to Develop Control Systems for Mobile Robots. Adaptive Behavior. 5,3/4, 343–363.

Nolfi, S. 1997b. Evolving Non-Trivial Behavior on Autonomous Robots: Adaptation is More Powerful Than decomposition and Integration. In Gomi (Ed.). Evolutionary Robotics. From Intelligent Robots to Artificial Life (ER'97). AAI Books.

Nolfi, S. 1997c. Evolving non-Trivial Behaviors on Real Robots: a garbage collecting robot. Robotics and Autonomous Systems. In Press.

Nolfi, S., Floreano, D., Miglino, O. and Mondada, F. 1994. How to evolve autonomous robots: Different approaches in evolutionary robotics. In Brooks and Maes (Eds.). Artificial Life IV. The MIT Press/Bradford Books.

Nolfi, S. and Parisi, D. 1993. Auto-teaching: Networks that develop their own teaching input. In Deneubourg, Bersini, Goss, Nicolis and Dagonnier (Eds.). Proceedings of the Second European Conference on Artificial Life. Free University of Brussels.

Nolfi, S. and Parisi, D. 1995. Evolving non-trivial behaviors on real robots: an autonomous robot that picks up objects. In Gori and Soda (Eds.). Topics in Artificial Intelligence. Proceedings of the Fourth Congress of the Italian Association for Artificial Intelligence. Springer.

Nolfi, S. and Parisi, D. 1997. Learning to Adapt to Changing Environments in Evolving Neural Networks. Adaptive Behavior, 5, 1, 75–98.

Nordin, P. and Banzhaf, W. 1996. An On-Line Method to Evolve Behavior and to Control a Miniature Robot in Real Time with Genetic Programming. Adaptive Behavior, 5, 2, 107–140.

Ram, A. and Arkin, R. and Boone, G. and Pearce, M. 1994. Using Genetic Algorithms to Learn Reactive Control Parameters for Autonomous Robot Navigation. Adaptive Behavior, 2(3), 277–305.

Salomon, R. 1996. Increasing Adaptivity through Evolution Strategies. In Maes, Mataric, Meyer, Pollack and Wilson (Eds.). Proceedings of the Fourth International Conference on Simulation of Adaptive behavior: From Animals to Animats 4. The MIT Press/Bradford Book.

Salter, M., Knowles, R. and Moncada, S. 1991. Widespread tissue distribution, species distribution and changes in activity of Ca^{2+}-dependent and Ca^{2+}-independent nitric oxide synthases. FEBS Lett.,291, 145–149.

Sanchez, E. and Tomassini, M. (Eds.). 1996. Towards Evolvable Hardware. The Evolutionary Engineering Approach. Springer Verlag.

Schwefel, H.P. 1995. Evolution and Optimum Seeking.Wiley.

Smith, T. 1997. Adding Vision to Khepera: An Autonomous Robot Footballer. Master's Thesis. School of Cognitive and Computing Sciences, University of Sussex.

Thompson, A. 1995. Evolving electronic robot controllers that exploit hardware resources. In Moran, Moreno, Merelo and Chacon (Eds.). Advances in Artificial Life: Proceedings of the Third European Conference on Artificial Life. Springer Verlag.

Thompson, A. 1997. Artificial Evolution in the Physical World. In Gomi (Ed.). Evolutionary Robotics. From Intelligent Robots to Artificial Life (ER'97). AAI Books.

Van Gelder, T. 1995. What Might Cognition Be If Not Computation. Journal of Philosophy, XCII(7), 345–381.

Wheeler, M. 1996. From Robots to Rothko: the bringing forth of worlds. In The Philosophy of Artificial Life, M. Boden (Ed.), OUP, 209–236.

Willshaw, D and Dayan, P. 1990. Optimal plasticity from matrix memories: What goes up must come down. Neural Computation, 2, 85–93.

Yamauchi, B. 1993. Dynamical neural networks for mobile robot control. Naval Research Laboratory Memorandum Report AIC-033-93. Washington.

Yamauchi, B. and Beer, R. 1994. Integrating Reactive, Sequential, and Learning Behavior Using Dynamical Neural Networks. In Cliff, Husbands, Meyer and Wilson (Eds.). Proceedings of the Third International Conference on Simulation of Adaptive behavior: From Animals to Animats 3. The MIT Press/Bradford Book.

How Co-evolution Can Enhance the Adaptive Power of Artificial Evolution: Implications for Evolutionary Robotics

Stefano Nolfi[1]   Dario Floreano[2]

[1]Institute of Psychology, National Research Council
Viale Marx 15, Roma, Italy
stefano@kant.irmkant.rm.cnr.it
[2]LAMI - Laboratory of Microcomputing
Swiss Federal Institute of Technology
EPFL, Lausanne, Switzerland
Dario.Floreano@epfl.ch

Abstract. Co-evolution (i.e. the evolution of two or more competing populations with coupled fitness) has several interesting features that may potentially enhance the power of adaptation of artificial evolution. In particular, as discussed by Dawkins and Krebs [2], competing populations may reciprocally drive one another to increasing levels of complexity by producing an evolutionary "arms race". In this paper we will investigate the role of co-evolution in the context of evolutionary robotics. In particular, we will try to understand in what conditions co-evolution can lead to "arms races" in which two populations reciprocally drive one another to increasing levels of complexity.

1. Introduction

Co-evolution (i.e. the evolution of two or more competing populations with coupled fitness) has several interesting features that may potentially enhance the adaptation power of artificial evolution. First, because the performance of the individual in a population depends also on the individual strategies of the other population which vary during the evolutionary process, the ability for which individuals are selected is more general (i.e. it has to cope with a variety of different cases) than in the case of an evolutionary process in which co-evolution is not involved. The generality of the selection criterion is a very important property because the more general the criterion, the larger the number of ways of satisfying it (at least partially) and the greater the probability that better and better solutions will be found by the evolutionary process.

Consider for example the well-studied case of two co-evolving populations of predators and prey [1]. If we ask the evolutionary process to catch one individual prey we may easily fail. In fact, if the prey is very efficient, the probability that an individual with a randomly generated genotype may be able to catch it is very low. As a consequence, all individuals will be scored with the same null value and the selective process cannot operate. On the contrary, if we ask the evolutionary process to find a predator able to catch a variety of different preys, it is much more probable that it will find an individual in the initial generations able to catch at least one of them and then select better and better individuals until one predator able to catch the original individual prey is selected.

Secondly, competing co-evolutionary systems are appealing because the ever-changing fitness landscape, due to changes in the co-evolving species is potentially useful in preventing stagnation in local minima. From this point of view, co-evolution may have consequences similar to evolving a single population in an ever-changing environment. Indeed the environment changes continuously given the fact that the co-evolving species is part of the environment of each evolving population.

Finally, the co-evolution of competing populations may produce increasingly complex evolving challenges. As discussed by Dawkins and Krebs [2] competing populations may reciprocally drive one another to increasing levels of complexity by producing an evolutionary "arms race". Let us again consider the predator and prey case: the success of predators implies a failure of the prey and conversely, when preys evolve to overcome the predators they also create a new challenge for them. Similarly, when the predators overcome the new preys by adapting to them, they create a new challenge for the preys. Clearly the continuation of this process may produce an ever-greater level of complexity. As Rosin and Belew [3] point out, it is like producing a *pedagogical* series of challenges that gradually increase the complexity of the corresponding solutions.

This nice property overcomes the problem that if we ask evolution to find a solution to a complex task we have a high probability of failure while if we ask evolution to find a solution first to a simple task and then for progressively more complex cases, we are more likely to succeed. Consider the predators and preys case again. At the beginning of the evolutionary process, the predator should be able to catch its prey which have a very simple behavior and are therefore easy to catch, likewise, prey should be able to escape simple predators. However, later on, both populations and their evolving challenges will become progressively more and more complex. Therefore, even if the selection criterion remains the same, the adaptation task will become progressively more complex and more general.

Unfortunately however a continuous increase in complexity is not guaranteed. In fact, co-evolving populations may cycle between alternative class of strategies that although they do not produce advantages in the long run may produce a temporary improvement over the co-evolving population. Imagine, for example, that in a particular moment population A adopts the strategy A_1 which gives population A an advantage over population B which adopts strategy B_1. Imagine now that there is a strategy B_2 (similar to B_1) that gives population B an advantage over strategy A_1. Population B will easily find and adopt strategy B_2. Imagine now that there is a strategy A_2 (similar to A_1) that provides an adaptive advantage over strategy B_2. Population A will easily find and adopt strategy A_2. Finally imagine that previously discovered strategy B_1 provides an advantage over strategy A_2. Population B will come back to strategy B_1. At this point also population A will come back to strategy A_1 (because, as explained above, it is effective against strategy B_1) and the cycle of the same strategies will be repeated over and over again (Fig. 1).

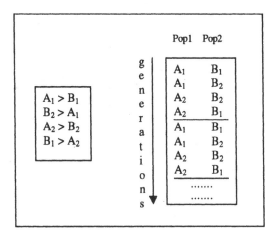

Fig. 1. The same strategies (A_1 and A_2 in population A) and (B_1 and B_2 in population B) may be selected over and over again throughout generations as is shown in the right hand side of the figure if the interaction between them looks like what is represented on the left side of the Figure. In this case the repeated cycle corresponds to 4 different combinations of strategies

Notice how the cycling may involve two or more different strategies for each population but also two or more different groups of strategies.

Of course this type of phenomena may cancel out all the previously described advantages because, although evolution may never get stuck in a particular solution, the number of different solutions discovered might be quite limited. Moreover there is no need to discover progressively more complex strategies. It is sufficient to re-discover previously selected strategies that can be obtained with a limited number of changes.

In this paper we will investigate the role of co-evolution in the context of evolutionary robotics. In particular, we will try to understand in what conditions co-evolution can lead to "arm races" in which two populations reciprocally drive one another to increasing levels of complexity.

2. Co-evolving Predator and Prey Robots

Several researchers have investigated co-evolution in the context of predators and prey in simulation [1, 4, 5, 6]. More recently, we have tried to investigate this framework first by using realistic simulations based on the Khepera [7, 8] and subsequently the real robots [9].

In this section, we will first describe our experimental framework and the results obtained in a simple case. Then, we will describe other two experimental conditions more suitable to the emergence of 'arm races' between the two competing populations.

2.1 The Experimental Framework

As often happens, predators and prey belong to different species with different sensory and motor characteristics. Thus, we employed two Khepera robots, one of which (the Predator) was equipped with a vision module while the other (the Prey) had a maximum available speed set to twice that of the predator. The prey has a black protuberance, which can be detected by the predator everywhere in the environment. The two species could evolve in a square arena 47 x 47 cm in size with high white walls so that predator could always see the prey (within the visual angle) as a black spot on a white background (see Fig. 2).

Both individuals were provided with eight infrared proximity sensors (six on the front side and two on the back) which had a maximum detection range of 3-4 cm in our environment. For the predator we considered the K213 module of Khepera which is an additional turret that can be plugged in directly on top of the basic platform. It consists of a 1D-array of 64 photoreceptors which provide a linear image composed of 64 pixels of 256 gray-levels each, subtending a view-angle of 36°. However the K231 module also allows detection of the position in the image corresponding to the pixel with minimal intensity. We exploited this facility by dividing the visual field into five sectors of about 5° each corresponding to five simulated photoreceptors. If the pixel with minimal intensity lay inside the first sector, then the first simulated photoreceptor would become active; if the pixel lay inside the second sector, then the second photoreceptor would become active, etc. From the motor point of view, we set the maximum wheel speed in each direction to 80mm/s for the predator and 160mm/s for the prey.

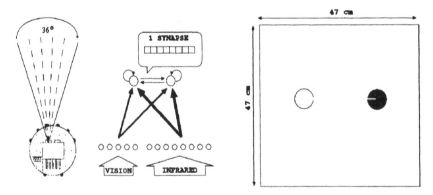

Fig. 2. Left and center: details of simulation of vision, of neural network architecture, and of genetic encoding. The prey differs from the predator in that it does not have 5 input units for vision. Eight bits code each synapse in the network. Right: Initial starting position for prey (left, empty disk with small opening corresponding to frontal direction) and predator (right, back disk with line corresponding to frontal direction) in the arena. For each competition, the initial orientation is random

In line with some of our previous work [10], the robot controller was a simple perceptron comprising two sigmoid units with recurrent connection at the output layer. The activation of each output unit was used to update the speed value of the corresponding wheel every 100ms. In the case of the predator, each output unit received connections from five photoreceptors and from eight infrared proximity sensors. In the case of the prey, each output unit received input only from 8 infrared proximity sensors, but its activation value was multiplied by 2 before setting the wheel speed. This structure, which is well-suited for the evolution of Braitenberg-like obstacle avoidance, was chosen as being a minimally sufficient architecture to evolve something interesting while maintaining system complexity at a manageable level; for the same reason, the architecture was kept fixed, and only synaptic strengths and output units threshold values were evolved.

In order to keep things as simple as possible and given the small size of the parameter set, we used direct genetic encoding [11]: each parameter (including recurrent connections and threshold values of output units) was encoded using 8 bits. Therefore, the genotype of the predator was 8 x (30 synapses + 2 thresholds) bits long while that of prey was 8 x (20 synapses + 2 thresholds) bits long.

Two populations of 100 individuals were each co-evolved for 100 generations. Each individual was tested against the best competitors of the previous generations (a similar procedure was used in [6, 12]). In order to improve co-evolutionary stability, each individual was tested against the best competitors of the ten previous generations (on this point see also below). At generation 0, competitors were randomly chosen within the same generation, whereas in the other 9 initial generations they were randomly chosen from the pool of available best individuals of previous generations.

For each competition, the prey and the predator were always positioned on a horizontal line in the middle of the environment at a distance corresponding to half the environment width, but always at a new random orientation. The competition ended either when the predator touched the prey or after 500 motor updates (corresponding to 50 seconds at maximum on the physical robot). The fitness function for each competition was simply 1 for the predator and 0 for the prey if the predator was able to catch the prey and, conversely 0 for the predator and 1 for the prey if the latter was able to escape the predator. Individuals were ranked after fitness performance in descending order and the best 20 were allowed to reproduce by generating 5 offspring each. Random mutation (bit substitution) was applied to each bit with a constant of probability pm=0.02[1].

For each set of experiments we ran 10 replications starting with different randomly assigned genotypes.

In this paper we will refer to data obtained in simulation. A simulator developed and extensively tested on Khepera by some of us [13] was used. However some of the experiments described have also been successfully replicated on real [9].

---

[1] The parameters used in the simulations described in this paper are mostly the same as in the simulation described in [7]. However, in these experiments we used a simpler fitness formula (a binary value instead of a continuous value proportional to the time necessary for the predator to catch the prey). Moreover, to keep the number of parameters as small as possible, we did not use crossover. In the previous experiments, in fact, we did not notice any significant difference in experiments conducted with different crossover rates.

2.2 Measuring Adaptive Progress in Co-evolving populations

In the co-evolutionary case, the Red Queen effect [14] makes it hard to monitor progress by taking measures of the fitness throughout generations. In fact, because fitnesses are defined relative to a co-evolving set of traits in the other individuals, the fitness landscapes for the co-evolving individuals vary. As a consequence, for instance, periods of stasis in the fitness value of the two populations may correspond to a period of tightly-coupled co-evolution.

To avoid this problem, different measure techniques have been proposed. Cliff and Miller [15] have devised a way of monitoring fitness performance by testing the performance of the best individual in each generation against all the best competing ancestors which they call CIAO data (Current Individual vs. Ancestral Opponents).

A variant of this measure technique has been proposed by some of us and has been called Master Tournament [7]. It consists in testing the performance of the best individual of each generation against each best competitor of all generations. This latter technique may be used to select the best solution from an optimization point of view [7]. Both techniques may be used to measure co-evolutionary progress (i.e. the discovery of more general and effective solutions).

2.3 Co-Evolution of Predator and Prey Robots: A Simple Case

The results obtained by running a set of experiments with the parameter described in Section 2.1 are shown below. Fig. 3 represents the results of the Master Tournament, i.e the performance of the best individual of each generation tested against all best competitors. The top graph represents the average result of 10 simulations. The bottom graph represents the result of the best run.

These results show that, at least in this case, phases in which both predators and preys produce increasingly better results are followed by sudden drops in performance. As a consequence, if we look at the average result of different replications in which increase and drop phases occur in different generations, we observe that performance does not increase at all throughout the generations. In other words the efficacy and generality of the different selected strategies does not increase evolutionarily. In fact, individuals of later generations do not necessarily score well against competitors of much earlier generations (see Fig. 4, right side). Similar cases have been described [3, 6].

The 'arm races' hypothesis would be verified if, by measuring the performance of each best individual against each best competitor, a picture approximating that shown on the left side of Fig. 4 could be obtained. In this ideal situation, the bottom-left part of the square, which corresponds to the cases in which predators belong to more recent generations than the prey, is black (i.e. the predator wins). Conversely, the top right part of the square, which corresponds to the cases in which the prey belong to more recent generations than the predators, is white (i.e. the prey wins). Unfortunately, what actually happens in a typical run is quite different (see right part of Fig. 4). The distribution of black and white spots does not differ significantly in the two sub-parts of the square.

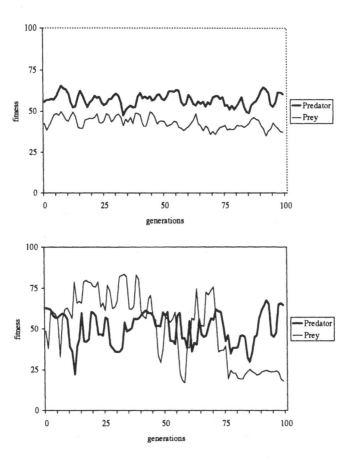

Fig. 3. Performance of the best individuals of each generation tested against all the best opponents of each generation (Master Tournament). Performance may range from 0 to 100 because each individual is tested once against each best competitor of 100 generations. The top graph shows the average result of 10 different replications. The bottom graph shows the result in the best replication (i.e. the simulation in which predators and preys of a given generations attain their best performance). Data were smoothed using rolling average over three data points

This does not imply that the co-evolutionary process is unable to find interesting solutions [7]. This merely means that effective strategies may be lost instead of being retained and refined. Such good strategies, in fact, are often replaced by other strategies that, although providing an advantage over the current opponents, are much less general and effective in the long run. In particular, this type of process may lead to the cycling process described in Section 1.2 in which the same strategies are lost and re-discovered over and over again.

The cycling between the same class of strategy is actually what happens in these experiments. If we analyze the behaviors of the best individuals of successive generations we see that in all replications, evolving predators discover and rediscover two different classes of strategies: (A_1) track the prey and try to catch it by approaching it; (A₂) track the prey while remaining more or less in the same area and attacking the prey only on very special occasions (when the prey is in a particular position relative to the predator). Similarly the prey cycles between two class of strategies: (B₁) stay still or hidden close to a wall waiting for the predator and eventually try to escape when the IR sensors detect the predator; (B₂) move fast in the environment, avoiding walls.

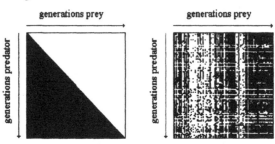

Fig. 4. Performance of the best individuals of each generation tested against all the best opponents of each generation. The black dots represent individual tournaments in which the predators win while the white dots represent tournaments in which the prey wins. The picture on the left represents an ideal situation in which predators are able to catch all prey of previous generations and the prey are able to escape all predators of previous generations. The picture on the right represents the result for the best simulation (the same shown in Fig. 3).

Now, as in Fig. 1, the strategy A_1 is generally effective against B_1, in fact the predator will reach the prey if the prey does not move too much and has a good chance to succeeding given that the prey can only detect predators approaching from certain directions. Strategy B_2 is effective against strategy A_1 because the prey is faster than the predator and so, if the predator tries to approach a moving fast prey, it has little chance of catching it. Strategy A_2 is effective against strategy B_2 because, if the prey moves fast in the environment, the predator may be able to catch it easily by waiting for the prey itself to come close to the predator. Finally, strategy B_1 is very effective against strategy A_2. In fact if the predator does not approach the prey and the prey stays still, the prey will never risk being caught. This type of relation between different strategies produces a cycling process similar to that described in Fig. 1.

What actually happens in the experiments is not so simple as in the description we have just given because of several factors: (1) the strategies described are not single strategies but classes of similar strategies. So for example there are plenty of different ways for the predator to approach the prey and different ways may have different probabilities of being successful against the same opposing strategies; (2) the advantage or disadvantage of each strategy against another strategy varies quantitatively and is probabilistic (each strategy has a given probability of beating a competing strategy); (3) populations at a particular generation do not include only one strategy but a certain number of different strategies although they tend to converge toward a single one; (4) different strategies may be easier to discover or re-discover than others.

However the cycling process between the different class of strategies described above can be clearly identified. By analyzing the behavior of the best individuals of the best simulation (the same as that described in Fig. 3 and 4), for example, we can see that the strategy B_2 discovered and adopted by preys at generation 21 and then abandoned after 15 generations is rediscovered and re-adopted at generation 58 and then at generation 98. Similarly the strategy A_2, first discovered and adopted by the predator at generation 10 and then abandoned after 28 generations for strategy A_1, is then rediscovered at generation 57. Interestingly, however, preys also discover a variation of strategy B_1 that includes also some of the characteristics of strategy B_2. In this case, preys move in circles waiting for the predator as in strategy B_1. However, as soon as they detect the predator with their IR sensors, they start to move quickly exploring the environment as in strategy B_2. This type of strategy may in principle be effective against both strategies A1 and A2. However sometimes preys detect the predator too late, especially when the predator approaches the prey from its left or right rear side which is not provided with IR sensors. Also, it might be that this hybrid strategy which is effective against both predator-strategies, it is not as effective against either predator strategy as the appropriate `pure' escape strategies. Therefore the hybrid strategy, despite its generalized effectiveness, is eventually turned into one of the pure strategies, namely the one that is more effective against whatever strategy is, at that time, being adopted by the predator.

2.4 Testing Individuals against All Discovered Solutions

In a recent article, Rosin and Belew [3], in order to encourage the emergence of 'arms races' in a co-evolutionary framework suggested saving and using as competitors all the best individuals of previous generations:

> *So, in competitive coevolution, we have two distinct reasons to save individuals. One reason is to contribute genetic material to future generations; this is important in any evolutionary algorithm. Selection serves this purpose. Elitism serves this purpose directly by making complete copies of top individuals.*
> *The second reason to save individuals is for purposes of testing. To ensure progress, we may want to save individuals for an arbitrarily long time and continue testing against them. To this end, we introduce the 'hall of fame', which extends elitism in time for purposes of testing. The best individual from every generation, is retained for future testing.*

From Rosin and Belew [3], pp. 8.

This type of solution is of course implausible from a biological point of view. Moreover, we may expect that, by adopting this technique, the effect of the co-evolutionary dynamic will be progressively reduced throughout generations with the increase in number of previous opponents. In fact, as the process goes on, there is less and less pressure to discover strategies that are effective against the opponent of the current generation and greater and greater pressure to develop solutions capable of improving performance against opponents of previous generations.

However, as the authors show, this method may be much more effective than a co-evolutionary framework in which individuals compete only with opponents of the

31

same generation. More specifically, we think, it may be a way to overcome the problem of the cycling of the same strategies. In this framework in fact, ad hoc solutions that compete successfully against the opponent of the current generation but do not generalize to opponents of previous generations cannot spread in evolving populations.

We applied the *hall of fame* selection regime to our predator and prey framework and measured the performance of each best individual against each best competitor (Master Tournament). As shown in Fig. 5 and 6, in this case, we obtain a progressive increase in performance. Results are obtained by running a new set of 10 simulations in which each individual is tested against 10 opponents randomly selected from all previous generations. All the other parameters remain the same.

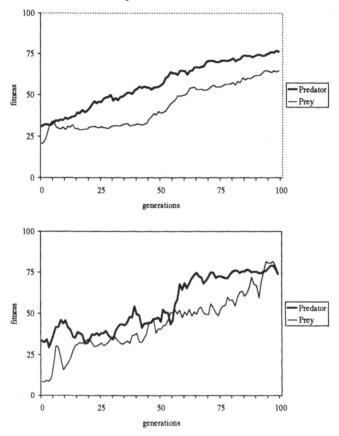

Fig. 5. Performance of the best individuals of each generation tested against all the best opponents of each generation (Master Tournament). The top graph shows the average result of 10 different replications. The bottom graph shows the result in the best replication (i.e. the simulation in which predators and prey of a given generation attain the best performance). Data were smoothed using a rolling average over three data points

Fig. 5 shows how in this case the average fitness of the best individuals tested against all best competitors progressively increases throughout generations, ultimately attaining near to optimal performances. Fig. 6 shows how this is accomplished by being able to beat most of the opponents of previous generations. The results do not exactly match the ideal situation described in Fig. 4 (left side) in which predators and prey are able to beat all individuals of previous generations. In the best simulation described in Fig. 5 (bottom graph) and Fig. 6, for example, there are two phases in which preys are unable to beat most of the predators of few generations before. The general picture, however, approximates the ideal one.

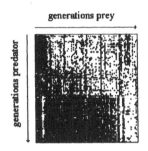

Fig. 6. Performance of the best individuals of each generation tested against all the best opponents of each generation. Black dots represent individual tournaments in which the predators win while white dots represent tournaments in which the prey wins. Result for the best simulation (the same shown in Fig. 5)

If we look at the strategies selected in this set of experiments we see that they are of the same class as those described in the previous Section. However, in this case the strategies are more stable (i.e. in general they are not suddenly replaced by another strategy of a different class). This enables the co-evolutionary process to progressively refine the current strategies instead of cycling between different classes of strategies restarting each time from about the same point.

2.5 How 'Arms Races' Can Continue Progressively to Produce More General Solutions in Certain Conditions

In section 2.3 we showed how 'arms races' spontaneously emerge in a co-evolutionary framework. However, we also showed how the innovations produced by such a process may be easily be lost because the evolutionary process tends to fail in a dynamic attractor in which the same type of solutions are adopted over and over by the two co-evolving populations. In section 2.4 we showed how the tendency to cycle between the same type of strategies may be reduced by preserving all previously discovered strategies and by using all of them to test the individual of the current population. However we also pointed out that this techniques which is biologically implausible, has its own drawbacks which may prevent it from scaling up.

In doing so, however, we also learned what characteristics may cause the sudden loss of the acquired abilities which often have to be rediscovered later on. As we showed in Section 2.3, evolution tends to produce the alternation of the same solutions over and over when there are two or more different classes of solutions that interact in a certain way among themselves. This implies that, if such conditions are not verified, 'arms races' should in principle be able to produce better and better solutions without falling into cycling periods.

Of course, it is not easy to predict the cases in which the conditions that produce cycling between the same strategies are absent. However, by analyzing the type of solutions selected by evolution in the experiments described above, we can try to make some predictions. One thing to consider, for example, is that the prey has a limited sensory system that enables it to perceive predators only at a very limited distance and not from all relative directions (there are no IR sensors able to detect predators approaching from the rear-left and rear-right side). Given this limitation, the prey cannot improve its strategy above a certain level. It can compete with co-evolving predators only by suddenly changing strategy as soon as predators select an effective strategy against them. However, if we increase the richness of the prey's sensory system we may expect that the prey will be able to overcome well adapted predators by refining its strategy instead of radically changing its behavior.

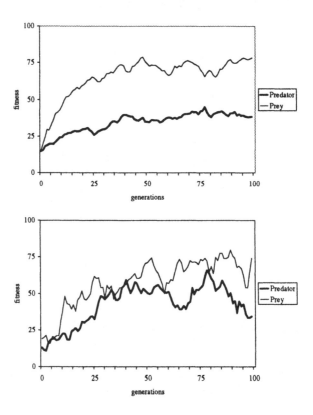

Fig. 7. Performance of the best individuals of each generation tested against all the best opponents of each generation (Master Tournament). The top graph shows the average result of 10 different replications. The bottom graph shows the result in the best replication (i.e. the simulation in which predators and prey of a given generation attain the best performance). Data were smoothed using a rolling average over three data points

To investigate this hypothesis we ran a new set of simulations in which the prey also was provided with a camera able to detect the predators' relative position. For the prey we considered another turret under development at LAMI, which consists of an 1D-array of 150 photoreceptors which provide a linear image composed of 150 pixels of 256 gray levels each [16]. We chose this wider camera because the prey, by escaping the predators, will only occasionally perceive opponents in their frontal direction. As, in the case of predators, the visual field was divided into five sectors of 44° corresponding to five simulated photoreceptors subtending a view-angle of 220°. As a consequence, in this experiment, both predator and prey are controlled by a neural network with 13 sensory neurons. Moreover, in this case, both predator and prey could see their competitors as a black spot against a white background. As in the experiments described in Section 2.3, individuals were tested against the best competitors of the 10 previous generations (not against competitors selected from all previous generations as in the experiments described in Section 2.4). All other parameters remained the same.

If we measure the average performance of the best predators and prey of each generation tested against all the best opponents of each generation (Master Tournament) a significant increase in performance throughout generations is observed (Fig. 7). In the case of the best replication, in particular, although predators' performance decrease in the last 20 generations, the best individuals up to generation 80 are able to overcome most of their opponents of previous generations (Fig. 8).

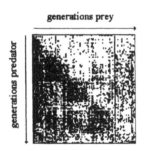

generations prey

generations predator

Fig. 8. Performance of the best individuals of each generation tested against all the best opponents of each generation. Black dots represent individual tournaments in which the predators win while white dots represent tournaments in which the prey wins. Result for the best simulation (the same as that shown in Fig. 7)

These results show how by changing the initial conditions (in this case by changing the sensory system of one population) 'arms races' can continue to produce better and better solutions in both populations without falling into cycles.

Interestingly, in their simulations in which also the sensory system of the two co-evolving populations was under evolution, Cliff and Miller observed that ".. pursuers usually evolved eyes on the front of their bodies (like cheetahs), while evaders usually evolved eyes pointing sideways or even backwards (like gazelles)." [16, pp.506]. The authors did not provide enough data in their paper to understand whether their simulations fell into solution cycles. However, even though both the nervous system and the sensory system were under co-evolution in their case, it seems that Cliff and

Miller did not observe any co-evolutionary progress toward increasingly general solutions. In fact, they report that 'co-evolution works to produce good pursuers and good evaders through a pure bootstrapping process, but both types are rather specially adapted to their opponents' current counter-strategies." [16, pp. 506]. However, it should be noted that there are several differences between Cliff and Miller experiments and ours. The fitness function used in their experiments, in fact, is more complex and includes additional constraints that try to force evolution in a certain direction (e.g. predators are scored for their ability to approach the prey and not only for their ability to catch it). Moreover, the genotype-phenotype mapping is much more complex in their cases and includes several additional parameters that may effect the results obtained.

3. Discussion

Evolutionary Robotics is a promising new approach to the development of mobile robots able to act quickly and robustly in real environments. One of the most interesting features of this approach is that it is a completely automatic process in which the intervention of the experimenter is practically limited to the specification of a criterion for evaluating the extent to which evolving individuals accomplish the desired task. However, it is still not clear how far this approach can scale up.

From this point of view, one difficult problem is constituted by the fact that the probability that one individual within the initial generations is able to accomplish the desired task, at least in part, is inversely proportional to the complexity of the task itself. As a consequence, if we apply this methodology to solving a complex task we are likely fail because all individuals of the initial generations are scored with the same zero values and as a consequence the selection mechanism cannot operate. We will refer to this problem as the *bootstrap problem.*

This problem arises from the fact that in artificial evolution people usually start from scratch (i.e. from individuals obtained with randomly generated genotypes). In fact, one possible solution to this problem is the use of 'incremental evolution'. In this case, we start with a simplified version of the task and, after we get individuals able to solve such a simple case, we progressively move to more and more complex cases [17, 18, 19]. This type of approach can overcome the bootstrap problem, although it also has the negative consequence of increasing the amount of supervision required and the risk of introducing inappropriate constraints. In the case of incremental evolution in fact, the experimenter should determine not only an evaluation criterion but also a 'pedagogical' list of simplified criteria. In addition the experimenter should decide when to change the selection criterion during the evolutionary process. Some of these problems may arise also when, although the selection criterion is left unchanged throughout the evolutionary process, it is designed to include rewards also for sub-components of the desired behavior [20].

Another possible solution of the bootstrap problem is the use of co-evolution. Co-evolution of competing populations, in fact, may produce increasingly complex evolving challenges spontaneously without any additional human intervention. Unfortunately however, no continuous increase in complexity is guaranteed. In fact, the co-evolutionary process tends to fail into dynamical attractors in which the same solutions are adopted by both populations over and over (we will refer to this problem as the *cycling problem*). What happens is that at a certain point one population, in order to overcome the other population, finds it more useful to suddenly change its strategy instead of continuing to refine it. This is usually followed by a similar rapid change of strategy in the other population. The overall results of this process is that most of the characters previously acquired are not appropriate in the new context and therefore are lost. However, later on, a similar sudden change may bring the two populations back to the original type of strategy so that the lost characters are probably rediscovered again and again.

The effect of the cycling problem may be reduced by preserving all the solutions previously discovered in order to test the individuals of the current generations [3]. However, this method has drawbacks that may affect some of the advantages of co-evolution. In fact, as the process goes on there is less and less pressure to discover strategies that are effective against the opponent of the current generation and increasing pressure to develop solutions able to improve performance against opponents of previous generations which are no longer under co-evolution.

We believe that the cycling problem, as the local minima problem in gradient-descent methods (i.e. the risk of getting trapped in a sub-optimal solution when all similar solutions produce a decrease in performance), is an intrinsic problem of co-evolution that cannot be eliminated completely. However, we also believe that the negative effects of such a problem do not apply to all cases and so completely cancel out the advantages of co-evolution. There may be cases, such as that described in Section 2.5, in which co-evolution may progressively produce more complex solutions for a large number of generations without losing the acquired characters by cycling between different types of solutions.

Moreover, different mechanisms may be able to limit the problems caused by the tendency to cycle between the same types of solutions.

Ontogenetic plasticity, for example, may allow individuals of one population to cope with different classes of strategies adopted by the second population, thus reducing the adaptive advantage of a sudden shift in the behavior causing the cycling problem (on the effects of some forms of ontogenetic plasticity in this framework see [8]). Interestingly, one can argue that co-evolution not only creates the adaptive pressure for ontogenetic adaptation (i.e. the ability to adapt during one's lifetime to different types of opponents' strategies produce a significant increase in the adaptation level of one individual) but also create the conditions in which ontogenetic adaptations can easily arise. In fact, coevolution, by falling into cycles of different classes of strategies, tends to select individuals which can shift from one class of strategy to another with only a few changes at the genotype level. It is plausible to argue that, for such individuals, a limited number of changes during ontogeny will be able to produce the required behavioral shift. In other words, it will be easier for co-evolving individuals to change their behavior during their lifetime to adopt strategies already adopted by their close ancestors thanks to the cycles occurring in previous generations.

Another factor that may limit the cycling problem is the richness of the environment. In the case of co-evolution, competing individuals are part of the environment. This means that part, but not all of the environment is undergoing co-evolution. Now the probability that a sudden shift in behavior will produce viable individuals is inversely proportional to the richness of the environment that is not under co-evolution. Imagine, for example, that an ability acquired under co-evolution, such as the ability to avoid inanimate obstacles, involves a characteristic of the environment which is not under co-evolution. In this case it will be less probable than a sudden shift in strategy involving the lost of such ability will be retained. In fact the acquired character will always have an adaptive value independently of the current strategies adopted by the co-evolving population. The same argument applies to the cases in which one population is co-evolving against more than one other population. The probability of retaining changes that involve a sudden shift in behavior will decrease because, in order to be retained, such changes would have to provide an advantage over both co-evolving populations.

References

1. Miller, G. F., Cliff, D.: Co-Evolution of Pursuit and Evasion I: Biological and Game-Theoretic Foundations, Technical Report CSRP311, School of Cognitive and Computing Sciences, University of Sussex (1994)
2. Dawkins, R., Krebs, J.R.: Arms races between and within species. Proceedings of the Royal Society of London B, **205** (1979) 489:511
3. Rosin, C.D., Belew, R.D.: New methods for competitive coevolution, Evolutionary Computation, **5** (1997) 1-29
4. Koza, J.R.: Evolution and co-evolution of computer programs to control independently-acting agents. In Meyer, J.A., Wilson, S. (eds.): From Animals to Animats: Proceeding of the First International Conference on Simulation of Adaptive Behavior, Cambridge, MA:MIT Press (1991)
5. Koza, J.R.: Genetic programming: On the programming of computers by means of natural selection, Cambridge, MA: MIT Press (1992)
6. Cliff, D., Miller, G.F.: Co-evolution of pursuit and evasion II: Simulation Methods and results. In Maes, P., Mataric, M., Meyer, J-A., Pollack, J., Roitblat, H., Wilson, S. (eds.): From Animals to Animats IV: Proceedings of the Fourth International Conference on Simulation of Adaptive Behavior, Cambridge, MA: MIT Press-Bradford Books (1996)
7. Floreano, D., Nolfi, S.: God Save the Red Queen! Competition in Co-Evolutionary Robotics. In Koza, J-R., Kalyanmoy, D., Dorigo, M., Fogel, D.B., Garzon, M., Iba, H., Riolo, R.L. (eds): Genetic Programming 1997: Proceedings of the Second Annual Conference, San Francisco, CA: Morgan Kaufmann (1997)
8. Floreano, D., Nolfi, S.: Adaptive behavior in competing co-evolving species. In Husband, P., Harvey, I. (eds): Proceedings of the Fourth European Conference on Artificial Life, Cambridge, MA: MIT Press (1997)
9. Floreano, D., Nolfi, S., Mondada, F.: Co-evolutionary Robotics: From Theory to Practice, Technical Report, EPFL, Swiss Federal Institute of Technology (1998)
10.Floreano, D., Mondada, F.: Automatic creation of an autonomous agent: Genetic evolution of a neural-network driven robot. In Cliff, D., Husband, F., Meyer J-A., Wilson, S. (eds.): From Animals to Animats III: Proceedings of the Third International Conference on Simulation of Adaptive Behavior, Cambridge, MA: MIT Press-Bradford Books (1994)
11.Yao, X.: A review of evolutionary artificial neural networks, International Journal of Intelligent Systems, **4** (1993) 203-222
12.Sims, K.: Evolving 3D morphology and behavior by competition. Artificial Life, **1** (1995) 353-372

13. Miglino, O., Lund, H. H., Nolfi, S.: Evolving Mobile Robots in Simulated and Real Environments. Artificial Life, **4** (1995) 417-434

14. Ridley, M.: The Red Queen: Sex and the evolution of human nature. Viking: London (1993)

15. Cliff, D., Miller, G.F.: Tracking the read queen: Measurement of adaptive progress in co-evolutionary simulations. In Moran, F., Moreno, A., Merelo, J.J., Chacon, P. (eds.): Advances in Artificial Life: Proceedings of the Third European Conference on Artificial Life, Berlin:Springer Verlag (1995)

16. Landolt, O.: Description et mise en oeuvre du chip ED084V2A, Technical Report 16-11-95, Centre Suisse d'Electronique et Microtechnique, Switzerland (1996)

17. Floreano, D.: Emergence of Home-Based Foraging Strategies in Ecosystems of Neural Networks. In: Meyer, J-A., Roitblat, H.L., Wilson, S. (eds.): From Animals to Animats II: Proceedings of the Second International Conference on Simulation of Adaptive Behavior, Cambridge, MA:MIT Press-Bradford Books (1993)

18. Harvey, I., Husbands, P., Cliff, D.: Seeing The Light: Artificial Evolution, Real Vision. In Cliff, D., Husbands, P., Meyer, J-A., Wilson, S. (eds.) From Animals to Animats III: Proceedings of the Third International Conference on Simulation of Adaptive Behavior, Cambridge, MA: MIT Press-Bradford Books (1994)

19. Gomez, F., Miikkulainem, R.: Incremental Evolution of Complex General Behavior, Adaptive Behavior, **5** (1997) 317-342

20. Nolfi, S. Evolving non-trivial behaviors on real robots: a garbage collecting robot, Robotics and Autonomous System 646, Special issue on "Robot learning: The new wave" (1997) 1-12

Running Across the Reality Gap: Octopod Locomotion Evolved in a Minimal Simulation

Nick Jakobi

School of Cognitive and Computing Sciences,
University of Sussex, Brighton, BN1 9QH, UK
nickja@cogs.susx.ac.uk

Abstract. This paper describes experiments in which neural network control architectures were evolved in minimal simulation for an octopod robot. The robot is around 30cm long and has 4 infra red sensors that point ahead and to the side, various bumpers and whiskers, and ten ambient light sensors positioned strategically around the body. Each of the robot's eight legs is controlled by two servo motors, one for movement in the horizontal plane, and one for movement in the vertical plane, which means that the robots motors have a total of sixteen degrees of freedom. The aim of the experiments was to evolve neural network control architectures that would allow the robot to wander around its environment avoiding objects using its infra-red sensors and backing away from objects that it hits with its bumpers. This is a hard behaviour to evolve when one considers that in order to achieve any sort of coherent movement the controller has to control not just one or two motors in a coordinated fashion but sixteen. Moreover it is an extremely difficult set-up to simulate using traditional techniques since the physical outcome of sixteen motor movements is rarely predictable in all but the simplest cases. The evolution of this behaviour in a minimal simulation, with perfect transference to reality, therefore, provides essential evidence that complex motor behaviours can be evolved in simulations built according to the theory and methodology of minimal simulations.

1   Introduction

Evolutionary Robotics is not magic and as several authors have pointed out [2, 6, 12, 14], there are many big questions that need answers if Evolutionary Robotics is to progress beyond the proof of concept stage. One of the most urgent of these (in that if it is not answered, Evolutionary Robotics is not going to progress very far at all) concerns how evolving controllers should best be evaluated. If they are tested using real robots in the real world, then this has to be done in real time, and the evolution of complex behaviours will take a prohibitively long time. If controllers are tested using simulations then the amount of modelling necessary to ensure that evolved controllers work on the real robot may mean that the simulation is so complex to design and so computationally expensive that all potential speed advantages over real-world evaluation are lost. How then should controllers be evaluated when testing in both simulation and reality seems fraught with insurmountable problems?

The author's recent thesis [8] offers an answer to this question. It does this by presenting new ways of thinking about and building simulations for the evaluation of evolving robot controllers. These *minimal simulations* run extremely fast and are trivially easy to build when compared to more conventional types of real-world simulation, yet they are still capable of evolving controllers for real robots. Thus the many advantages of using simulations are preserved while most of the major disadvantages are avoided. The aim of this paper is to show how complex motor behaviours can be evolved in minimal simulation. Space limitations do not allow a full explication of the theory and methodology of minimal simulations. However, the next section gives a broad brush sketch of the main ideas, which, together with the details of the minimal simulation used for the experiments reported in this paper, should give the reader a good idea of how to build and use a minimal simulation.

Fig. 1. The Octopod robot.

This paper describes experiments in which neural network control architectures were evolved for an octopod robot. The robot, shown in figure 1 is around 30cm long and has 4 infra red sensors that point ahead and to the side, various bumpers and whiskers, and ten ambient light sensors positioned strategically around the body. Each of the robot's eight legs is controlled by two servo motors, one for movement in the horizontal plane, and one for movement in the vertical plane, which means that the robots motors have a total of sixteen degrees of freedom.

The aim of the experiments was to evolve neural network control architectures that would allow the robot to wander around its environment avoiding objects using its infra-red sensors and backing away from objects that it hits with its bumpers. This is a hard behaviour to evolve when one considers that in order to achieve any sort of coherent movement the controller has to control not just one or two motors in a coordinated fashion but sixteen. Moreover it is an extremely difficult set-up to simulate using traditional techniques since the physical outcome of sixteen motor movements is rarely predictable in all but the simplest cases. The evolution of this behaviour in a minimal simulation, therefore, provides essential evidence that complex motor behaviours can be evolved in simulations built according to the theory and methodology put forward in [8, 9, 7]. See [3, 4, 11] for related work on evolving locomotion controllers for walking robots and for abstract computer models of insects.

The minimal simulation used to evolve controllers for the octopod is described in Section 3. The rest of the evolutionary machinery, including the neural networks, the encoding scheme, the genetic algorithm and genetic operators is described in section 4. Experimental results are put forward in section 5 and finally, in section 6, some comments are offered on the paper as a whole. But first, a brief description of the idea of minimal simulations.

2 Minimal simulations

The artificial evolution of controllers typically involves the constant and repetitive testing of hundreds upon thousands of individuals as to their ability to behave in a certain way or perform a certain task. In the case of real robots this testing procedure is far from a trivial matter and (with the exception of certain hybrid approaches [15, 14]) can be done in only one of two ways: controllers must either be evaluated on real robots in the real world, or they must be evaluated in simulations of real robots in the real world. Both of these approaches have their problems.

As [12] point out, the evaluation of controllers on real robots must be done in real time, and this probably makes the entire evolutionary process prohibitively slow. But even if we are resigned to an evolutionary process that takes years rather than days, then there are different problems that must be faced. The process must be automated, for instance. This begs many technological questions to do with power supplies, wear and tear, automatic fitness evaluations ans so on.

As has been shown by several experimenters [10, 1, 13], it *is* possible to evolve controllers in simulation for a real robot. Now that this is no longer in doubt the question becomes one of whether the technique will scale up. In [12] (and similar points were made earlier in [6, 2, 5]), the authors argue that if behavioural transference can only be guaranteed when a carefully constructed empirically validated simulation is used, then as robots and the behaviours we want to evolve for them become more complicated, so do the simulations. The level of complexity involved, they argue, would make such simulations:

- so computationally expensive that all speed advantages over real-world evolution are lost.
- so hard to design that the time taken in development outweighs time saved by fast evolution.

Clearly the main challenge for the simulation approach to evolutionary robotics is to invent a general theoretical and methodological framework that enables the easy and cheap construction of fast-running simulators for evolving real-world robot behaviours. This is what is provided in [8, 9, 7] where the general framework is developed and a wide range of examples are given in which evolved behaviours successfully crossed the reality gap (the evolved controllers performed perfectly on the real robots). The basic idea of a minimal simulation is to model only those robot-environment interactions that are necessary to underpin a desired behaviour. Everything else is made unreliable by careful use of randomness. In this way an evolved controller is forced to use the minimal set of interactions picked out by the simulation designer. This set of interactions is modelled as simply (i.e. computationally cheaply) as possible by using an envelope of noise to mask the inaccuracies of the modelling. Again, by careful use of randomness, controllers evolve that are robust to these inaccuracies and hence will cross the reality gap.

The whole methodology can be summarised in the following step-by-step guide to building a minimal simulation:

1. **Precisely define the behaviour.** Start by making a precise definition of the behaviour to be evolved. This should include both a description of the task to be performed by the robot(s) and the range of environmental conditions it is to be performed under.

2. **Identify the real-world base set.** Distinguish between those real-world features and processes that are relevant to the performance of the behaviour and those that are not. Those that are relevant constitute the base set. If possible, identify the way in which the members of the base set interact with each other and react to motor signals during the performance of the behaviour.

3. **Build a model of the way in which the members of the base set interact with each other and react to controller output (when the robot is performing the behaviour).** Make a model of the real-world base set of features and processes. The dynamics of this model need copy those of the real world only during the performance of successful behaviour. The dynamics when the behaviour is not being performed may often be shaped to smooth the fitness landscape, thus facilitating the evolution of successful controllers.

4. **Build a model of (enough of) the way in which the members of the base set affect controller input (when the robot is performing the behaviour).** Identify and model processes by which members of the real-world base set give rise to aspects of controller input. Just as with the model of the base set, these modelled processes need only copy their real-world counterparts when the behaviour is being performed. Make sure that the input aspects that these processes give rise to in the minimal simulation are sufficient to underly successful behaviour. Note that there are often several ways in which sufficient input processes may be identified and modelled, and the exact choice affects the possible strategies that evolving controllers may employ.

5. **Design a suitable fitness test.** Design a suitable fitness test that only awards maximum fitness points to those controllers that reliably perform the behaviour.

In particular, each evaluation must involve a sufficient number of trials so that, with the right amount of trial to trial variation (see below), we can be confident that controllers which achieve high fitness in all of them are reliably fit, base set exclusive and base set robust.

6. **Ensure that evolving controllers are base set exclusive.** Make a distinction between the implementation aspects (those features of the simulation that are there for coherence etc but which we do not want to underpin the desired behaviour) of controllers' input signals and the base set aspects. Those implementation aspects that are present during the performance of the behaviour must be randomly varied from trial to trial so that evolving controllers that depend upon them are unreliable. In particular, enough variation must be included to ensure that evolving controllers can not, in practice, be reliably fit unless they are base set exclusive i.e. they actively ignore each implementation aspect and depend exclusively on the base set aspects of their input to perform the behaviour.

7. **Ensure that evolving controllers are base set robust.** Every base set aspect of the simulation must be randomly varied from trial to trial so that reliably fit controllers are forced to be base set robust. The extent of this random variation must be large enough that controllers which evolve to be reliably fit are also able to cope with the inevitable differences between the base set aspects of the simulation and their real-world counterparts. Care should be taken that this variation is not so large that reliably fit controllers fail to evolve at all.

3   The Octopod minimal simulation

According to received wisdom, simulating something as complex from an actuator point of view as an eight-legged robot is hard. The problems arise from the fact that sixteen motors all moving at the same time and interacting with each other in the real world rarely induce movement in the robot that is easy to model and often produce completely unpredictable movement that is best looked at as stochastic. What happens when two legs clash, for instance? Or when the belly of the robot is on the ground but the legs attempt to push the robot forwards? Or when 4 of the legs attempt to push the robot forwards and 4 of the legs attempt to push the robot backwards? Clearly any simulation that sets out to model *all* of the dynamics of the system will involve vast quantities of pain-staking empirical measurement and research into friction-coefficients, the power of each motor, the range of possible movement of the robot and so on. If the only simulation in which we could evolve autonomous walking behaviour for the real robot was of this type then the simulation would be so complicated that it might indeed be simpler to evolve controllers on the real thing.

Happily we do not need to come close to modelling all of the possible dynamics of the robot in order to build a satisfactory minimal simulation. The key is to realise that those portions of the possible dynamics of an octopod robot which are difficult and complicated to model (the vast majority) are precisely those that are *not* involved in successful walking behaviour. When the octopod robot walks around its environment in an acceptable manner, its legs do *not* clash and its belly does *not* drag along the ground and its legs do *not* pull in different directions. The minimal simulation described below takes full advantage of this fact. The dynamics of the simulated robot match the dynamics of the real robot only when the controller is inducing acceptable, successful walking and obstacle-avoiding behaviour. If a controller does anything else *but* acceptable, successful walking and obstacle-avoiding behaviour then the simulation falls woefully short of modelling what would actually happen in the real world. Since a controller that performs the behaviour will never take the robot into this region of the dynamics, we do not need to model it. The headings for the step-by-step guide to minimal simulations given in section 2 will be followed in describing the octopod simulation in detail.

Precisely define the behaviour.

The aim of the experiments was to evolve octopod-controllers that could walk around the environment, turning away from objects that fell within range of the IR sensors and backing away from objects that touched the front bumpers and whiskers. At the very least, this requires that controllers are able to perform 4 sub-behaviours, each relevant to a particular sensory scenario:

- If an object falls within range of the left-hand IR sensors then the robot must turn on the spot to the right.
- If an object falls within range of the right-hand IR sensors then the robot must turn on the spot to the left.
- If an object hits the front bumpers or whiskers then the robot must walk backwards as fast as possible.
- In the absence of any objects falling within infra-red range or touching the robot's front bumpers or whiskers, the robot must walk forwards in a straight line as fast as possible.

In a cluttered environment there are occasions in which other sensory combinations may occur e.g. objects may fall within range of the left and right IR sensors at the same time. However, these occasions are rare enough in simple environments to grant that controllers which are able to perform each of these 4 simple sub-behaviours are also capable of wandering around their environment satisfactorily without bumping into anything or becoming stuck.

One reason for making this behavioural reduction is that constructing a fitness test that specifically checks for each of the 4 sub-behaviours, one after the other, is actually much easier than constructing a fitness test that checks directly for the more complex global behaviour. We do not need to simulate, for example, the way in which the robot's position within a complex environment gives rise to sensor values. Instead we may test directly for each of the 4 sub-behaviours in turn by clamping the sensor values to fit each of the 4 sensory scenarios and observing the movement of the robot in response. The fitness function was therefore divided into 4 phases: each testing for one of the 4 behaviours outlined above.

The order in which each of the 4 phases occurred was random and evolving neural network controllers were not reset in between. This ensured that reliably fit controllers would be able to perform each of the 4 sub-behaviours independently.

Identify the real-world base set

Whether or not the robot satisfactorily performs each of the 4 sub-behaviours is a function of the movement of the robot body in each of the 4 different sensory scenarios. The members of the base set, therefore, are those features of the world that make up the causal pathway from controller output to how the body as a whole moves in response. These include the way in which controller output affects how the legs move, and the way in which the movement of the legs affects the movement of the body as a whole.

Build a model of the way in which the members of the base set interact with each other and react to controller output (when the robot is performing the behaviour).

The overall movement of the robot was described by two variables: one for the speed of the left-hand side of the robot and one for the speed of the right-hand side of the robot. Thus if both sides of the robot moved straight ahead at the same speed then the overall movement of the robot was deemed to be straight ahead, if they moved in different directions but with equal velocity then the robot was deemed to be turning on the spot, and if both sides moved backwards at the same speed then the overall movement of the robot was deemed to be straight backwards.

To model the way in which the robot as a whole moved in response to controller output, therefore, necessitated a model of the way in which each leg responded to controller output, and the way in which the movement of each leg contributed to the overall movement of each *side* of the robot. However, because of the arguments put forwards in [8], it was not necessary to accurately model the way in which *every* motor signal could affect the movement of the robot as a whole, but only those motor signals involved in satisfactory walking forwards, backwards and turning on the spot. The dynamics of the model, therefore, matched those of reality only for those controllers that prevented the body from touching the ground, moved all the legs supporting the robot on each side in the same direction (either all forwards or all backwards depending on whether the robot was supposed to be walking forwards, backwards or turning on the spot), and kept those legs that were not touching the ground as high in the air as possible.

The motor signals to the servo-motors controlling the legs of the octopod robot specify absolute angular positions (relative to the body) that the servo-motors are required to move the legs to. Thus when a new signal is sent to the servo-motor controlling the horizontal or vertical angle of a particular leg, it will move as fast as possible to the new location. In the absence of any new signal, the leg will remain rigid. This process was modelled in the simulation

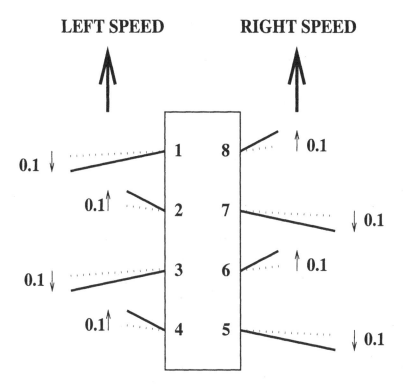

Fig. 2. This figure shows diagrammatically how the speeds of the left-hand and right-hand sides of the robot were calculated from the vertical and horizontal positions of the eight legs. For explanatory purposes the length of each leg in the diagram is inversely proportional to its height above the ground so that the long legs are 0.8 as low as they can go and the short legs are 0.2 as low as they can go. Adding up the contributions that each leg makes to the speed of its side we see that the speeds of both the left and the right hand side of the robot work out at $0.1 \times 0.8 - 0.1 \times 0.2 + 0.1 \times 0.8 - 0.1 \times 0.2 = 0.12$ forwards

by calculating, on every iteration, horizontal and vertical angular displacements for each leg based on the differences between the angular positions specified by the motor signals and the actual angular positions of the simulated legs. The maximum possible angular speed of each leg was measured very roughly and set in the simulation to be 2π radians per second. Using the horizontal and vertical angles of each leg, a simple look up table provided the approximate position, relative to the robot, that each leg projected onto the ground, and the 4 legs in the lowest positions were assigned as the supporting legs. A simple calculation was then made to see whether the robot's centre of gravity was contained within the polygon subtended by the floor-contact positions of these 4 legs, in which case the robot was deemed to be stable. If it was not, then the robot was deemed to be unstable. Also the average height of these 4 legs relative to the robot body was calculated. If they were low enough then the robot was deemed to be standing, otherwise it was deemed to be dragging its belly on the ground.

Figure 2 shows diagrammatically how the speeds of the left and right-hand sides of the robot, and thus the overall movement of the robot, were calculated from the controller's motor signals. On each iteration, the contribution each leg made to the forwards or backwards movement of its side of the robot was worked out according to a simple calculation. The distance moved by the leg (either forwards or backwards) was multiplied by a figure between 0 and 1 that was inversely proportional to how high in the air the leg was. Thus the higher in the air a leg was, the smaller the contribution its horizontal movement made to the total movement of its side of the robot. The nearer to the ground the leg was, the larger the contribution its horizontal movement made to the total movement of its side of the robot. The contributions that each leg makes were then added up to arrive at a figure for the total movement (either forwards or backwards) of that side of the robot. If both the left and the right side of the robot moved forwards then the robot was deemed to have moved forwards, if both sides moved backwards then the robot was deemed to have moved backwards, if each side moved in different directions then the robot was deemed to be turning on the spot.

Now although this simple model seems to bear no relationship to reality (how can a leg that is in the air contribute to the speed of its side of the robot?), a controller that made maximum use of the dynamics of the model to move the robot around as fast as possible would keep all of the legs that were moving in the wrong direction at any one time as high in the air as possible and all the legs that were moving in the appropriate direction as firmly on the ground as possible. Since penalty terms for both instability and belly-dragging were included in the fitness function (see below), maximally fit controllers remained stable and stood upright at all times, moving all the legs that were supporting the robot on each side in the same direction (either all forwards or all backwards depending on whether the robot was walking forwards, backwards or turning on the spot) and keeping those legs that were not supporting the robot as high in the air as possible.

Build a model of (enough of) the way in which the members of the base set affect controller input (when the robot is performing the behaviour).

The sensor model employed was so simple as to be almost non-existent. The sensors were divided into three groups: the front left and back left IR sensors forming one group, the front right and back right IR sensors forming another group, and the front whiskers and bumpers forming another. In the phase of each fitness test in which there were no objects within sensor range, all sensors were set to background levels for the duration of the phase: 0 for the bumpers and whiskers and 255 for the IR sensors. In the phases during which an object fell within IR range on either the left or right-hand side of the robot, the IR sensor on the appropriate side was set to high (200) for the duration of the phase. In the phase during which an object hit the touch sensors, the front whiskers and bumper were set to high (1), *but only for the first second of the phase*. This simple sensor model provided evolving controllers with enough information about the world to perform the behaviour satisfactorily.

Design a suitable fitness test

As explained above, each fitness evaluation was divided into 4 phases: one for each of the 4 sensory scenarios. Each of these phases lasted five simulated seconds. At the end of every iteration of the simulation, the fitness of the controller being tested was incremented by a value δ derived from the overall movement of the robot. How this value was calculated depended on the sensory scenario the robot was in at the time:

- If there were no objects within sensor range then δ was the speed of the left-hand side of the robot plus the speed of the right-hand side of the robot.
- If there was an object within infra-red sensor range on the right-hand side of the robot then δ was the speed of the right-hand side of the robot *minus* the speed of the left-hand side of the robot.
- If there was an object within infra-red sensor range on the left-hand side of the robot then δ was the speed of the left-hand side of the robot *minus* the speed of the right-hand side of the robot.
- If an object hit the bumpers then δ (for the duration of this phase of the fitness evaluation) was *minus* the speed of the left-hand side of the robot *minus* the speed of the right-hand side of the robot.

Also on each iteration, if the robot was deemed to be unstable then a small penalty was subtracted from the fitness, and if the robot was deemed to be touching the ground with its belly then a small penalty was subtracted from the fitness.

Ensure that evolving controllers are base set robust and base set exclusive

The fitness test described above was carefully designed so that controllers that evolved to be reliably fit would use only those portions of the simulation dynamics that corresponded closely to the dynamics of the real robot. In fact, these dynamics turned out to be close enough that there was no need to vary the simulation at all in order to ensure that evolved controllers were base set robust. Any differences between simulation and reality were easily accommodated by slop in the definition of satisfactory walking and obstacle-avoiding behaviour. Thus walking behaviour on the real robot might be a little jerkier or quicker than in the simulation, but it was still perfectly adequate walking behaviour.

Likewise, nothing extra was added to the simulation in order to ensure that evolving controllers were base set exclusive. This was for the simple reason that the model of the way in which sensor values arose from the base set was so simple that there was nothing else in the simulation that evolving controllers could come to rely upon.

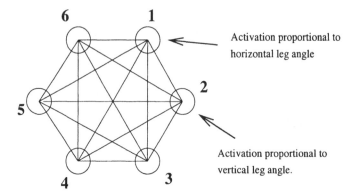

Fig. 3. Each leg controller consisted of six fully connected neurons. The activity of neuron 1 and neuron 2 controlled the horizontal and vertical leg angles respectively.

4 The evolutionary machinery

In this section we describe the evolutionary machinery that, together with the minimal simulation described above, was responsible for evolving neural network control architectures that could perform the behaviour satisfactorily in reality.

Neural networks

While network parameters (connection weights, time constants, thresholds and so on) were under evolutionary control in the experiments described in this section, the overall shape of the network architecture was fixed to be the same for every member of the population. The repetitive movements characteristic of multi-legged walking behaviours were produced by a main oscillatory network of 8 coupled sub-networks, each responsible for the direct control of a single leg. The properties of this oscillatory network were then modulated by the output from three sensory neurons (one each for left and right infra red and one for the bumpers) and one permanently saturated bias neuron to produce the different movement patterns for walking forwards, backwards and turning. This architecture is very similar to, and was based upon, that used by Gallagher and Beer in [1]. The components of this architecture will now be explained in detail.

Figure 3 shows one of the basic sub-networks responsible for the control of each leg. All eight sub-networks were identical in that only one set of sub-network parameters (threshold constants, connection weights and so on) was encoded on the genome and repeated eight times. These sub-networks consisted of six fully interconnected neurons, numbered 1 to 6 in the diagram, of the same type as those used by Gallagher and Beer in [1]. At each iteration, the input activity A_j of each of the $j = 1$ *to* 6 neurons in each of the 8 sub-networks was calculated according to equation 1

$$\tau_j \dot{A}_j = -A_j + \sum w_{ij} O_i + I_j \tag{1}$$

where τ_j was a time constant that affected the rate and extent to which the jth neuron responded to input, O_i was the output from the i_{th} neuron, w_{ij} was the weight on the connection from the i_{th} neuron to the j_{th} neuron, and I_j was any external input to the j_{th} neuron from outside the network. Once the input activity of each neuron had been calculated, the output O_j of each of the $j = 1$ to 6 neurons in each of the 8 sub-networks was calculated from the input activity A_j according to the sigmoid function of equation 2

$$O_j = (1 - e^{(t_j - A_j)})^{-1} \tag{2}$$

where t_j was a threshold constant associated with the j_{th} neuron. The range of possible values of each of these genetically specified constants is listed below.

The output of neuron 1 in each sub-network was responsible for the signal to the servo-motor controlling the horizontal motion of the leg in question, and the output of neuron 2 was responsible for the signal to the servo-motor that controlled the vertical angle of the leg (see figure 3). For neuron 1, an output of 0 mapped onto a signal to the horizontal servo motor to point as far backwards as it could go, and an output of 1.0 mapped onto a signal to the the servo motor to point the leg as far forward as it could go. For neuron 2, outputs of 1 and 0 mapped onto signals to the vertical servo motor to position the leg in the fully up and down positions respectively.

Each sub-network was coupled to the sub-network directly opposite it and to the network on either side of it (with wraparound) as in figure 4. Each sub-network to sub-network coupling involved six symmetrical connections: from neuron 1 in one network to neuron 1 in the other, from neuron 2 to neuron 2, neuron 3 to neuron 3 and so on. All 4 cross-body couplings were identical in the sense that only six connection strengths were encoded on the genome and this set of six was repeated 4 times. All 8 along-body couplings were identical in the same way.

Figure 5 shows an example of how the connections between the neurons that made up the leg-controller sub-networks could be modulated by the sensor neurons and the bias neuron. Each connection between leg-controller neurons can be thought of as having had a synapse half way down its length that acted as a gate: open and the connection was unaffected, closed and the connection was switched off, effectively reducing the weight on the connection to zero. The synapse itself received input from sensor neurons and the bias neuron by way of weighted connections. If the total input to the synapse was greater than zero then the synapse gate was open and the connection between the leg-controller neurons was unaffected. If the total input to the synapse was less than zero then the synapse gate was closed and the weight on the connection between the leg-controller neurons dropped to zero.

Figure 6 shows how the three sensor neurons and the bias neuron were connected up to the synapses of the leg-controller sub-networks. Each of the thick

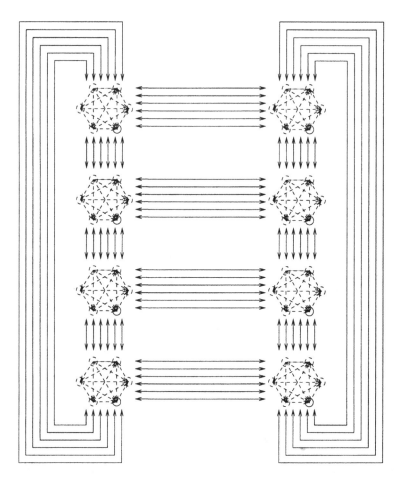

Fig. 4. This diagram shows how each leg-controller sub-network was coupled to the sub-network opposite it on the body and to the sub-network either side of it with wraparound.

black arrows represents 36 connections, one for each of the 36 synapses of a leg-controller sub-network. In total, three sets of 36 connection weights were encoded on the genome: one set for the infra-red sensor neurons, one for the bumper sensor neuron and one for the bias neuron. Thus each of the two infra-red sensor neurons were connected to the synapses of the leg-controller sub-networks on the appropriate side by way of four identical sets of 36 connections (both sets of four were also identical to each other), and both the bumper sensor neuron and the bias neuron were connected up to all eight leg-controller sub-networks by way of eight identical sets of connections each.

A weighted input connection was associated with each of the three sensor neurons and the bias neuron. The signals from the infra-red sensors and bumper

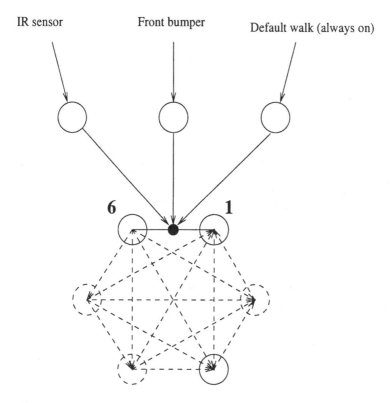

Fig. 5. This diagram shows how each connection between leg-controller neurons contains a synapse 'gate' that can be turned on or off by sensor neurons and the bias neuron. For the sake of diagrammatic simplicity only one connection is shown, whereas in reality every connection in every leg controller sub-network ($36 \times 8 = 288$ connections in total) contains a synapse that can be modulated in this way.

sensors that fed into these connections were normalised to lie within the range 0 to 1. In the case of the bias neuron, the signal that fed into its weighted input connection was permanently set at 1.

The network was updated iteratively using time-slicing techniques at a rate of 16 updates per second (or the simulated equivalent of a second). Also, in order to reduce computational overheads, a 200 place look-up-table was provided for the sigmoid function in place of the standard C-library maths functions.

Encoding scheme

Since the layout of the neural network architecture was fixed and predefined for every individual, a simple direct encoding scheme was employed. Every parameter was encoded on the genotype by a real number in the range 0 to 99, and this was mapped onto the relevant range during decoding. The parameters that

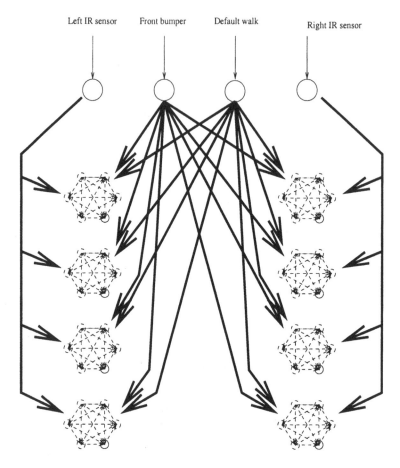

Fig. 6. This diagram shows how the sensory neurons and the always-on bias neuron were connected to the synapses of the leg controller sub-networks. Each of the thick black arrows represents 36 connections, one to each synapse in the leg controller. The cross-body and along-body couplings between leg-controller sub-networks have not been shown in this diagram.

were encoded and the ranges onto which they were mapped are as follows:

- 36 connection weights for the leg-controller sub-networks mapped onto the range ±16.
- 12 cross-body and along-body coupling connection weights mapped onto the range ±16.
- 36 infra-red sensor neuron to synapse connection weights mapped onto the range −6.5 to 25.5.
- 36 bumper sensor neuron to synapse connection weights mapped onto the range −6.5 to 25.5.

- 36 bias neuron to synapse connection weights mapped onto the range -6.5 to 25.5.
- 9 unit threshold constants mapped onto the range ± 4: 6 for the leg-controller sub-network neurons, 1 for the infra-red sensor neurons, 1 for the bumper sensor neuron and 1 for the bias neuron.
- 9 unit time constants mapped onto the range 0.5 to 5.0: 6 for the leg-controller sub-network neurons, 1 for the infra-red sensor neurons, 1 for the bumper sensor neuron and 1 for the bias neuron.
- 3 input connection weights mapped onto the range ± 16: 1 for the infra-red sensor neurons, 1 for the bumper sensor neuron and one for the bias neuron.

which makes a total of 177 parameters. Thus genotypes were strings of 177 numbers in the range 0 to 99.

Genetic algorithm and genetic operators

The genetic algorithm was an extremely simple generational model with tournament selection and elitism. After evaluating every member of the population, offspring genotypes were repeatedly produced until the next generation was full. To make a new offspring, two pairs of individuals were picked at random from the population and the fittest individuals from each pair (i.e. the winners of the tournaments) were chosen to act as parents. The offspring genotype was then formed from these two parents through a process of crossover and mutation: single point crossover was applied with a probability of 1, and every one of the 177 numbers that made up the offspring had a 0.02 chance of being mutated. A mutation involved changing the number in question by a random amount taken from a roughly normal distribution with a standard deviation of around 18. If the new value was greater than 99 or less that 0 then it was clipped to lie within this range.

5 Experimental results

After removing some initial bugs from the code[1], reliably fit controllers evolved on practically every run within around 3500 generations. This took around 14 hours to run on a Sun Ultra SPARC and simulated over 11 weeks worth of real world evolution. When downloaded onto the real octopod, reliably fit controllers made the robot walk around its environment in a satisfactory manner, turning away from objects that fell within infra red range on both the right and the left hand side and backing away from objects that it hit with its bumpers.

Unfortunately, we must make do with the bald statement of fact that evolved controllers successfully crossed the reality gap. It is not possible to demonstrate

---

[1] One such bug, spotted by Jerome Kodjabachian, meant that the penalty due to robot instability was effectively applied at random. Surprisingly, even with such a fundamental error in the code, controllers evolved that were able to perform the task perfectly satisfactorily when downloaded onto the robot.

it here due to both the nature of the octopod robot itself and the format of the paper. If the robot was equipped with position sensors on each of the legs then data recorded from these sensors as the robot moved around a real-world environment could be used to provide such a demonstration. The robot, however, is not equipped with sensors of this type and data of the required type is not available. The other form such a demonstration could take, and probably the most natural, is the evidence provided by video footage of the robot wandering around its environment. This cannot, however, be profitably presented as part of a text and pictures document; even if a sequence of stills taken at short and regular time intervals were displayed, this would not be all that informative as to how the legs of the robot moved in the real world unless there were an impractically large number of them.

In lieu of any method of demonstrating how the legs of the real robot moved as it wandered around its environment, the best we can do is to provide a demonstration of how the motor signal patterns to these legs change in response to each of the four sensory scenarios. Figure 7 offers such a demonstration for a typical reliably fit controller that evolved after 3200 generations. From top to bottom, the first eight traces provide a novel representation of the motor signals issued to each leg over the course of an average fitness trial, and the bottom two traces show the resultant velocities of the left and right side of the simulated robot respectively. The best way of explaining how to read the slightly bizarre looking motor traces is to describe how they were generated. At each iteration of the simulation, a short line representing the current motor signal was added to the right hand side of each motor signal trace. As can be seen from the figure, these lines were of various thicknesses and were always drawn from the horizontal centre line of the trace either up and to the left or down and to the left with various different gradients. The thickness of each line represented the vertical angle of the leg relative to the ground as specified by the motor signal in question: the thicker the line, the lower the leg, and the thinner the line the higher the leg. The gradient of the line represented the horizontal angle of the leg relative to the body as specified by the motor signal in question: the further up and to the left, the further forwards relative to the body, and the further down and to the left, the further backwards relative to the body. In this way, although they are perhaps harder to read than other less informative types of trace devised to convey similar information (see [1] for example), each of the motor signal traces of figure 7 represents *both* the vertical and horizontal components of the relevant signal over the course of a fitness test.

From the left and right velocity traces in figure 7 it is evident that the octopod moved forwards, then turned on the spot to the right, then backed up for a period and then rotated on the spot to the left. This corresponds to the order in which the four sensory scenarios arose during the fitness test that gave rise to this figure: no sensors active, left IR sensor active, bumpers and whiskers active, right IR active. Close inspection of the eight motor signal traces reveals:

- In the absence of any sensory activity the robot proceeded forwards using the classic tripod gate. Note that each leg is perfectly out of sync with the leg directly opposite it on the other side of the body.

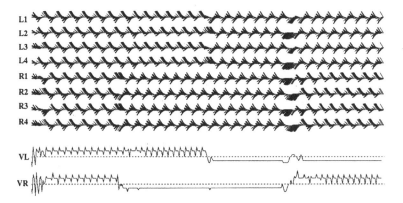

Fig. 7. Each leg controller consisted of six fully connected neurons. The activity of neuron 1 and neuron 2 controlled the horizontal and vertical leg angles respectively.

- In response to activity from either of the two IR sensors, the motor signals sent to each of the legs on the side of the robot furthest from the sensor suddenly became the exact opposite of the signals sent to each of the corresponding legs on the side of the robot nearest the sensor. This made the side nearest the sensor signal go forwards and the one furthest away move backwards.
- In response to activity from the bumpers and whiskers, the robot proceeded backwards using a backwards tripod gate. Note that just before this phase of the simulation was finished, but well after the short-lived inputs to bumpers and whiskers had ceased, the robot paused with all legs down and back for a moment.

When downloaded onto the real robot, these motor patterns and walking gates were clearly and reliably recognizable.

6 Comments

The minimal simulation used in this paper makes full use of the arguments put forward in [8] to evolve controllers for the octopod robot. Simply put, these arguments state that a minimal simulation need only model the real-world dynamics involved in successful behaviour and no others. This is because the only controllers that must cross the reality gap, if the simulation is to be a success, are precisely those that use these dynamics (i.e. perform the behaviour) and no others. For many robotics setups and behaviours this may not be of any use since the dynamics involved in successful behaviour may be neither obvious ahead of time nor qualitatively different to the rest of the dynamics of the system. For the experiments reported in this article, however, the dynamics of the octopod robot during successful walking and obstacle avoiding behaviour were both relatively easy to identify and *much* easier to model than the dynamics of the octopod robot as a whole. A minimal simulation that modelled these dynamics alone was therefore easy to construct and ran extremely fast when compared to the simulation that would result from attempting to model *all* of the dynamics of the octopod robot within its environment.

Acknowledgements

First, infinitely many thanks to Phil Husbands for editing this paper out of the thesis chapters I left him while I disappeared on the piste. What a fine man. Blame the title on him. Second, many thanks to Phil Husbands, Inman Harvey, Mike Wheeler, Giles Mayley, Joe Faith, Adam Bockrath, Seth Bullock, Jon Bird, Pete de Bourcier, Emmet Spier, Adrian Thompson, Suzy Levy and other members of COGS and the CCNR, past and present, for crucial discussions, debate and help. Third, a big salute to the Brighton beach 4:00am naked winter swimming club, I couldn't have done this without you. This work was supported by a COGS postgraduate bursary.

References

1. R.D. Beer and J.C. Gallagher. Evolving dynamic neural networks for adaptive behavior. *Adaptive Behavior*, 1:91–122, 1992.
2. R. Brooks. Artificial life and real robots. In F. J. Varela and P. Bourgine, editors, *Toward a Practice of Autonomous Systems: Proceedings of the first European Conference on Artificial Life*, pages 3–10, Cambridge, Massachusetts, 1992. MIT Press / Bradford Books.
3. F. Gruau. Automatic definition of modular neural networks. *Adaptive Behavior*, 3(2):151–184, 1995.
4. F. Gruau. Cellular encoding for interative evolutionary robotics. In P. Husbands and I. Harvey, editors, *Fourth European Conference on Artificial Life*, pages 368–377. MIT Press/Bradford Books, 1997.
5. I. Harvey and P. Husbands. Evolutionary robotics. In *Proceedings of IEE Colloquium on 'Genetic Algorithms for Control Systems Engineering', London 8 May 1992*, 1992.
6. P. Husbands and I. Harvey. Evolution versus design: Controlling autonomous robots. In *Integrating Perception, Planning and Action, Proceedings of 3rd Annual Conference on Artificial Intelligence, Simulation and Planning*, pages 139–146. IEEE Press, 1992.
7. N. Jakobi. Half-baked, ad-hoc and noisy: Minimal simulations for evolutionary robotics. In P. Husbands and I. Harvey, editors, *Fourth European Conference on Artificial Life*, pages 348–357. MIT Press/Bradford Books, 1997.
8. N. Jakobi. *Minimal Simulations for Evolutionary Robotics*. PhD thesis, University of Sussex, 1998.
9. N. Jakobi. Evolutionary robotics and the radical envelope of noise hypothesis. *Adaptive Behavior*, 6(2), 1998, pages =.

10. N. Jakobi, P. Husbands, and I. Harvey. Noise and the reality gap: The use of simulation in evolutionary robotics. In F. Moran, A. Moreno, J.J. Merelo, and P. Chacon, editors, *Advances in Artificial Life: Proc. 3rd European Conference on Artificial Life*, pages 704–720. Springer-Verlag, Lecture Notes in Artificial Intelligence 929, 1995.

11. J. Kodjabachian and J.-A. Meyer. Evolution and development of neural networks controlling locomotion, gradient following and obstacle avoidance in artificial insects. *IEEE Transactions on Neural Networks*, page (in press), 1998.

12. M.J. Mataric and D. Cliff. Challenges in evolving controllers for physical robots. *Robot and Autonomous Systems*, 19(1):67–83, 1996.

13. O. Miglino, H.H. Lund, and S. Nolfi. Evolving mobile robots in simulated and real environments. *Artifical Life*, 2(4), 1995.

14. S. Nolfi, D. Floreano, O. Miglino, and F. Mondada. How to evolve autonomous robots: Different approaches in evolutionary robotics. In R. Brooks and P. Maes, editors, *Artificial Life IV*, pages 190–197. MIT Press/Bradford Books, 1994.

15. A. Thompson. Evolving electronic robot controllers that exploit hardware resources. In F. Moran, A. Moreno, J.J. Merelo, and P. Chacon, editors, *Advances in Artificial Life: Proc. 3rd European Conference on Artificial Life*, pages 640–656. Springer-Verlag, Lecture Notes in Artificial Intelligence 929, 1995.

Detour Behavior in Evolving Robots: Are Internal Representations Necessary?

Orazio Miglino[1], Daniele Denaro[2], Guido Tascini[3], and Domenico Parisi[4]

[1] II University of Naples, Via G.Paolo I, 81055 S.Maria.C.V., Italy
orazio@caio.irmkant.rm.cnr.it
[2] Department of Computer Science,University of Ancona, Via Brecce Bianche, 60131
Ancona, Italy
denaro@caio.irmkant.rm.cnr.it
[3] Department of Computer Science,University of Ancona, Via Brecce Bianche, 60131
Ancona, Italy
tascini@inform.unian.it
[4] Institute of Psychology, National Research Council, Via Marx 15, 00137 Rome, Italy
domenico@kant.irmkant.rm.cnr.it

Abstract. Internal representations of the environment are often invoked to explain performance in tasks in which an organism must make a detour around an obstacle to reach a target and the organism can lose sight of the target along the path to the target. By simulating a detour task in evolving populations of robots (Khepera) we show that neural networks with memory units perform better than networks without memory units in this task. However, the content of the memory units need not be interpreted as an internal representation of the position of target. The memory units send a time-varying internally generated input to the network's hidden units that allows the network to generate the appropriate behavior even when there is no external input. Networks without memory units do not have this internal input and this explains their inferior performance.

1   Detour behavior

Imagine a robot (an artificial organism) that must reach a target object located somewhere in the environment. The robot has a camera mounted on its body with which the robot can see the target. Since the camera has a restricted visual field the robot turns until the target happens to be within its visual field and it then can move in the direction of the target and reach it.

If there is no obstacle in the environment the task is easy enough (cf. e.g., [5]). Consider however an environment that includes an obstacle in addition to the target object. The obstacle is a low wall that allows the robot to see the target even if the target is located on the other side of the wall (provided the robot is oriented in the direction of the target) but prevents the robot from proceeding directly toward the target. The only solution for the robot is to make a detour around the wall in order to reach the target. But this creates a serious problem for our robot. When the robot is making a detour around the wall the target is likely to exit the robot's visual field. The problem is particularly serious if the wall has the shape shown in Figure 3b. In order to negotiate the obstacle the robot may be forced to go in the opposite direction with respect to the target and therefore to loose sight of the target. (The type of task we have described is called "detour task" and is a standard procedure for testing real organisms' spatial abilities. [3, 9])

An organism which is able to make a detour around an obstacle is sometimes said to possess an "internal representation" of the spatial environment that mediates between stimulus and response (cf., e.g., [8]). When the organism sees the target and it can solve the task by simply approaching the target (i.e., there is no obstacle), the organism can directly map stimulus into response. However, when the organism must make a detour around an obstacle and therefore it is forced to lose sight of the target, one can think the organism is in possession of some internal representation of the environment that tells the organism where is the target even if the target is currently not visually accessible.

We do not find the notion of an internal representation particularly perspicuous, although this notion is often invoked in explaining the spatial behavior of organisms [6, 10]. In the present paper we analyze the detour behavior of (artificial) organisms by using a neural network model of the nervous system controlling the robot's behavior. We train a population of networks using a genetic algorithms so that evolved networks come to possess connection weights allowing them to to exhibit various degrees of the ability to make a detour around an obstacle in order to reach a target object. We compare two conditions. In one condition the robot's behavior is controlled by a simple feedforward neural network. The network has two distinct sets of input units encoding the input, respectively, from a camera and from a set of infrared sensors. The first set of input units (camera) encode the position of the target with respect to the robot provided the robot's body is turned in the direction of the target and therefore the target falls inside the robot's visual field; otherwise, the target is not perceived by the robot. The second set of input units (infrared sensors) encode the position of the robot's body with respect to either the obstacle or the peripheral wall enclosing the environment provided the body is sufficiently close to either the obstacle or the peripheral wall; otherwise, the robot does not perceive the obstacle or the enclosing wall. Both sets of input units send their activations to a layer of hidden units that in turn send their activations to two output units. The two output units encode the rotation speed of the two wheels that allow the robot to turn and move in space.

In the second condition the neural network includes an additional layer of special units called memory units [4]. In any particular cycle the activation level of each hidden unit is copied in the activation level of a corresponding memory unit. Each memory unit is linked with normal excitatory or inhibitory connections to all hidden units. Therefore in the next cycle the activation level of the hidden units - and therefore the network's output - depends both on the current activation of the input units (and the input-to-hidden weights) and on their own activation in the preceding cycle stored in the memory units (and the memory-to-hidden weights). The memory units maintain a cumulative trace of the past and make this trace available to the hidden units so that the network in each cycle generates an output influenced by both the current input from the environment and this trace of the past.

In choosing to compare these two conditions our reasoning is the following. (For other simulations exploring the role of memory in robot's or animat's behavior, see

[1, 2]). It is tempting to equate the existence of memory units in a network architecture with the possession of an internal representation of the environment that makes it possible for an organism to respond to features of the environment even when these features are not currently perceived. When a robot without memory units must make a detour around an obstacle and it loses sight of the target, it has no way of knowing where is the target. Therefore, it will have trouble orienting itself and regaining visual access to the target in order to approach and reach the target. On the contrary, if the neural network controlling the robot's behavior includes an additional layer of memory units, the problem can be more easily solved. If in cycle N the robot sees the obstacle and in cycle N+1 it looses sight of the obstacle, the trace of the location of the obstacle will remain somehow stored in the memory units. Therefore, even if the robot has no direct visual access to the target it can respond to the current input using this stored trace of the position of the target. This trace will allow the robot to orient itself and to turn towards the target, thereby regaining visual access to it.

What we would like to determine is whether a robot without memory units can solve the detour problem or, in order to solve the problem, it is necessary to have memory units that allow the robot to maintain a trace of the past. In the latter case the content of the memory units would constitute some sort of representation of the environment which is necessary to solve the detour problem. Another possibility is that both robots with memory units and robots without memory units can solve the detour problem but the robots with memory units perform better than those without memory units. As we will see, this second result is obtained in our simulations. However, we don't think that this means that the content of memory units constitutes an internal representation of the environment. As we will show, the role of the memory units in enhancing performance in a detour task can be analyzed in a simpler and more basic way that does not require internal representations

2. Simulations

In our simulations we use a small robot called Khepera which has a cylindrical body with a diameter of 5 cm and a height of 8 cm (Figure 1). Around the body there are 8 proximity (infrared) sensors with a nonlinear range of 3 cm. However, we use only 3 of these 8 sensors: left, front, and right. Mounted on the top of the robot there is a camera with a linear array of 64 pixels. The camera has a horizontal visual field of 36 degrees at 8 cm from the ground. Each pixel can have one of 256 grey levels. For our purposes the 64 pixels were reduced to 8 (sampled each 8 pixels on the entire array of 64 pixels) with only 2 grey levels (black and white). The robot turns and moves by regulating the rotation speed of two wheels.

62

Fig. 1. The Khepera robot

The environment is a rectangular walled arena of 55x40 cm. It contains a low obstacle that can have various shapes and a cyclindrical target. The proximity sensors of the robot can sense the obstacle or the target or the peripheral wall provided the robot is sufficiently close to them. The camera can see the target (but not the obstacle or the wall) at any distance (provided the target falls within its restricted visual field) even when the target is on the other side of the low obstacle with respect to the robot.

We developed a simulator for the robot and the environment because we wanted to use a genetic algorithm to train the robot and using a genetic algorithm with real robots is too expensive in terms of time and management. The simulator is based on a software library in C language for managing bidimensional environments and mobile units developed at our Institute. After a population of robots has evolved the required capacity we test the control systems (neural networks) in a real (physical) environment with the physical Khepera. Generally, very similar behaviors were exhibited by the simulated and the physical robots.

The control system of our robots is constituted by a neural network. We used two types of networks in two different populations: with and without memory. The network without memory is a normal feedforward neural network with a single layer of hidden units. The network has 11 input units (3 for the infrared sensors and 8 for the camera), 2 output units (for the robot's 2 wheels), and 5 hidden units. The network with memory has an additional layer of memory units, one for each hidden unit. In any given cycle each of the memory units is assigned the same activation level of the corresponding hidden unit. Since all the memory units are linked with normal connections to all the hidden units, in any given input/output cycle the activation pattern appearing on the hidden units (and therefore the network's output) depends on both the current activation pattern of the input units and the cumulative trace of the past stored in the memory units. (Notice that in the networks with memory the hidden units are 4, that is, one unit less than in the networks without memory. This guarantees a comparable total number of connections for the two types of networks (83 connections for the networks without memory units and and 89 connections for the networks with memory units) and therefore a connection space of comparable size to be searched by the genetic algorithm).

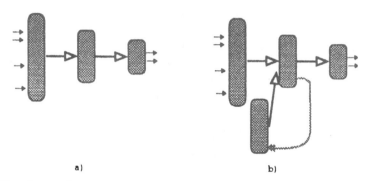

Fig. 2. Neural network controlling the robot's behavior: (a) without memory; (b) with memory

The simulation starts with a population of 100 organisms all with the same network architecture and randomly assigned connection weights. The weights are randomly selected from a rectangular distribution ranging from -5 to +5. Each individual lives alone in an environment that contains a target and an obstacle. The individual must make a detour around the obstacle to reach the target. The individuals that perform better are selected for reproduction. In order to facilitate the evolution of the detour behavior the population evolves in a simpler environment in the first 200 generations, and then for 200 additional generations in a slightly more complex environment. For the first 200 generations the environment contains a rectangular obstacle (a bar of 20 cm of length and 3 cm of height) and a target (a cylinder of 2 cm of diameter and 10 cm of height) (Figure 3a). The life of all individuals lasts a total of 240 input/output cycles divided into two epochs of 120 cycles each. At the beginning of each of the two epochs the individual is positioned in the environment below the obstacle facing North. The target is positioned near the upper left corner of the environment in one epoch and near the upper right corner in the other epoch in such a way that the target falls outside the organism's restricted visual field. The organism has available 120 cycles to make a detour around the obstacle and reach the target.

After the first 200 generations the environment changes and it becomes more complex. The obstacle now has a U-inverted shape (Figure 3b) and is moved somewhat more North. The lifetime of an organism lasts 6 epochs of 120 cycles each. In three of these 6 epochs the organism is initially placed below the obstacle facing North and the target is located once in the upper left corner, once in the upper right corner, and once midway between the two corners, i.e., in the middle of the upper portion of the environment (cf. Figure 3b). In the remaining 3 epochs there is no obstacle and the organism and the target are placed as in the 3 epochs with the obstacle. These 3 epochs with no obstacle were included to allow the organisms to evolve an ability to choose between a direct route to the target when no obstacle is present and an indirect detour route when there is an obstacle.

The fitness on the basis of which individuals are selected for reproduction is based on two components. An individual is assigned one fitness unit for each cycle the individual spends near the target, where nearness to the target means to be located 12 cm or less from the target. This is the first component. The second component is calculated according to the following formula (used in standard evolutionary robotics experiments [7]):

$$F = abs(V)*(1-sqrt(DV))*(1-I) \qquad (1)$$

Where V is velocity (absolute value of the sum of activations of the two wheels), sqrt(DV) is the square root of the difference between the activations of the two wheels, and I is the average activation of the 3 infrared sensors. In other words, the second fitness component rewards the individual if it moves fast and straight (without turning) and it punishes it for being too close to the obstacle or wall (i.e., when the infrared sensors are activated; notice that the second fitness component is ignored when the organism is near the target). An individual's performance is evaluated in each cycle using the above formula. An individual that exhibits the best possible performance is assigned one fitness unit. Individuals that are not so good are assigned less than one unit of fitness according to their performance.

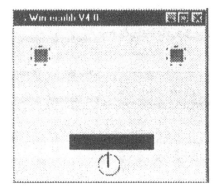 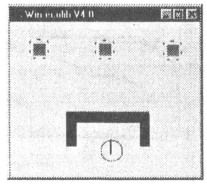

Fig. 3. First environment (a); second environment (b).The dotted areas identifie alternative target positions.

At the end of life each individual is assigned a fitness which is the average of the fitnesses obtained in each epoch of life (2 epochs in the first block of 200 generations and 6 epochs in the second block of 200 generations). The 100 individuals of each generation are ranked on the basis of their fitness and the 20 best individuals are selected for reproduction. Each of these 20 individuals generates 5 offspring, i.e., that is, 5 neural networks with the same connection weights of their (single) parent except that 20% of the weights are randomly modified by adding a quantity randomly selected in the interval between -1 and +1 to their current value.

3 Results

Figure 4 shows how fitness changes across the 200+200=400 generations for (a) the best individual, (b) the 20 best individuals (average), and (c) the total population of 100 individuals (average) in each generation, for both the population with memory and the population without memory. (The data are the average of 5 replications of the simulation with different "seeds" for constructing the initial population.) The fitness increases across the 400 generations, which implies that our organisms learn (in an evolutionary sense) the task, i.e., they acquire an ability to reach the target after making a detour around the obstacle. In most of the curves there is a rapid increase in fitness after generation 200 which appears to be due to the fact that in the second environment our organisms live half of their life (3 out of 6 epochs) in an environment without the obstacle, which allows them to reach the target more easily. Aside from that, there is a clear and significant advantage of the individuals with memory units in their network architecture over the individuals without memory units.

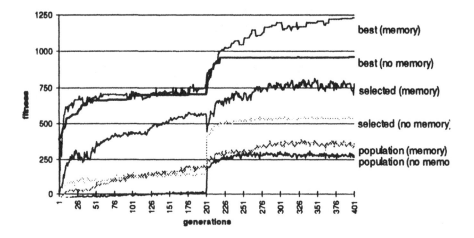

Fig. 4. Change in fitness across 400 generations for the best individual, the 20 best individuals, and the entire population of 100 individuals in each generation, for both the population withour memory and the population with memory.

At the end of evolution we tested for generalization the 5 best individuals of the last generation in each of the 5 replications of the simulation for both the population with memory and the population without memory, for a total of 25+25=50 individuals. Each of these 50 individuals is allowed to live for 100 epochs (120 cycles each) and in each epoch (a) the initial position of the organism is the same as during evolution but its orientation is chosen at random (whereas it was always North during evolution), (b) the location of the target in the upper portion of the environment is chosen at random (whereas it was restricted during evolution, cf. Figure 3b), and (c) the obstacle in different epochs is either (1) the bar of the first environment, (2) the inverted-U obstacle of the second environment, (3) an L-shaped obstacle, or (4) is absent. The performance in this generalization test (generalization test A) was measured as the percentage of epochs in which an organism was able to reach the target. An organism was considered as having reached the target if it was able to reach a location at less than 3 cm from the target.

The results indicate that when tested in an environment which is different from the environment they have experienced during evolution the organisms with memory perform more efficiently than the organisms without memory. They are able to reach the target in 78% of the epochs compared with 65% of the organism without memory. Using the t Student test we found a statistically significant difference ($t(48)=2.89$, $p<0.01$).

These results establish that keeping a neural trace of the past in such a way that in any given cycle the organism responds to both the current input and this trace of the past results in a better performance in the detour task both in the environment in which the task has been learned (has evolved) and in new environments never experienced before. Both the organisms with memory and the organisms without memory evolve an ability to make a detour around an obstacle to reach a target but the organisms with memory perform better than those without memory.

Then our next questions are: How is the behavior of the organisms with memory different from the behavior of the organisms without memory? How can the memory units help an organism to solve the task more efficiently?

To answer these questions we divide the entire behavioral sequence that takes the organism from its starting position to the target into two successive segments. In the first segment the organism must negotiate the obstacle and turn around it. In the second segment the organism leaves the obstacle behind itself and approaches the target. Operationally, the first behavioral segment ends when the organism finds itself North of the obstacle for the first time and the second behavioral segment includes everything which happens after that event. We will show that the organisms with memory and the organisms without memory behave differently with respect to both the first and second segments of the behavioral sequence.

As the fitness curves of Figure 4 show, an interesting result of our simulations is that even the organisms without memory are able to turn around the obstacle and head towards the target even if in order to do so they must necessarily lose sight of the target. These organisms can solve the problem by adopting the strategy of following the obstacle. They might approach the obstacle and respond to the sensory (infrared) input from the obstacle to generate movements that allow them to walk along the obstacle until they have turned around the obstacle and can regain visual access to the target. At this point they can simply approach the target. Or they can go away from the obstacle, approach the peripheral wall, and use the sensory (infrared) input from the wall to reach the upper portion of the environment where the target is located.

On the other hand, the organisms with memory have an alternative option open to them. When there is no visual input from the target because the organism is facing away from the target and there is no input from the infrared sensors because the organism is too distant from either the obstacle or the wall, the organism is completely deprived of sensory input since it is exposed to a constant zero input. The organisms without memory can only escape from this difficult situation by approaching either the obstacle or the wall in order to get some useful sensory input. The organisms with memory, on the contrary, can use the constantly changing internal input from their memory units to generate appropriate behaviors.

This analysis predicts that the organisms without memory will evolve behavioral trajectories that tend to bring them near either the obstacle or the peripheral wall more often than the organisms with memory. These trajectories allow the organisms without memory to be exposed to sensory input from either the obstacle or the wall more often than the organisms with memory. The organisms with memory do not need to do so because they have a time-varying input which is internally generated by the memory units.

To test this prediction we have determined the percentage of cycles in generalization test A (in the epochs with an obstacle) in which the organism receives sensory input from its infrared sensors, for both the organisms with memory and the organisms without memory. This percentage is 42% for the organisms with memory and 63% for the organisms without memory (the difference is statistically significant, $t(48)=3.19$, $p<0.01$). We conclude that the two types of organisms behave differently in the first portion of the behavioral sequence that allows them to reach the target. They both must turn around the obstacle and therefore lose sight of the target. However, their behavior of negotiating the obstacle is different. The organisms without memory tend to stay close to the obstacle or to approach the peripheral wall in order to receive sensory input from the obstacle or the wall that they can use to go around the obstacle and reach the upper portion of the environment where they will find the target. In fact, this input is the only input that can guide their behavior in the first segment of the behavioral sequence that will take them to the target. In contrast, the organisms with memory are able to turn around the obstacle and reach the upper portion of the environment with less input from their infrared sensors because they can count on the internal input from their memory units to guide their behavior.

We now turn to the second segment. The second segment begins when the organism has already made a detour around the obstacle and has reached the upper portion of the environment where the target is located. What is demanded of the organism at this stage is to turn toward the target in order to gain visual access to the target and to use this visual input to head toward the target. This is a simple stimulus-response task that can be solved without memory units and, therefore, at least in principle the behavior of the organisms with memory and the behavior of the organisms without memory should not be different at this stage.

To test how the two kinds of organisms behave in the second segment of the behavioral sequence we compare the behavior of the 25 individuals with memory and of the 25 individuals without memory in 100 epochs in which the environment contains no obstacle (generalization test B). At the beginning of each epoch (120 cycles) the organism is placed in the standard starting position at the center of the environment with a randomly selected orientation and the target is placed in a randomly selected position in the upper portion of the environment. We measure (a) the percentage of epochs in which an organism is able to reach the target (i.e., to reach a location at 3 or fewer centimeters from the target), (b) the average number of cycles necessary for an organism to reach the target, and (c) the percentage of cycles an organism receives sensory input from the wall.

The results show that there are no differences between the organisms with and without memory with respect to measure (a). Both types of organisms reach the target equally often, that is, in most of the epochs (92% for the organisms with memory; 90% for the organisms without memory). (Remember that there is no obstacle in the environment in this test.) However, the average time taken by the organisms with memory to reach the target is 20 cycles while it is 32 cycles for the organisms without memory (statistically significant difference; $t(48)=4.32$, $p<0.01$). Therefore, memory allows our organisms to reach the target more quickly. Furthermore, the organisms without memory are less likely than those without memory to end up near the peripheral wall in their search for the target. This is suggested by the fact that the average number of cycles in which the organisms with memory receive input from their infrared sensors is 40 (in each epoch) while it is 67 for the organisms without memory ($t(48)=2.97$, $p<0.01$).

Table 1 summarizes the results of the generalization tests. Typical trajectories of organisms with and without memory in an environment with an obstacle and in an environment with no obstacle are shown in Figure 5.

Table 1. Summary of generalization test results.

Type of measure	Without memory	With memory
% of target reachings in generalization test A	65	78
% of target reachings in generalization test B	90	92
# of cycles to reach the target in generalization test B	32	20
# of cycles to reach the target from first detection (test B)	21	10
# of cycles with infrared sensory input activated (test A)	63	42
# of cycles with infrared sensory input activated (test B)	67	40

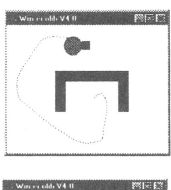
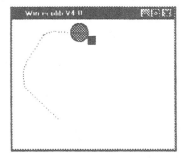
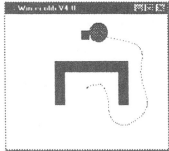

Fig. 5. Typical trajectories of organisms without memory (top) and with memory (bottom) in an environment with an obstacle (left column) and in an environment with no obstacle (right column).

4   Discussion

One important result of our simulations is that even simple input-output networks without memory units can to some extent solve a detour task. Although in some cycles the robot is necessarily unable to see the target because it has to turn away from the target in order to make a detour around the obstacle, the robot does not need to wander randomly in the environment until it happens to have gone past the obstacle and to regain visual access to the target. The robot can approach the obstacle and/or the peripheral wall and using the sensory input from its infrared sensors it can generate a trajectory that makes a detour around the obstacle and takes the robot to the upper portion of the environment where the target can be directly reached.

However, solving a detour problem with a simple feedforward neural network without memory units generates an inferior performance compared with solving the problem with memory units. The organisms with memory units perform better than those without memory units both during evolution and in generalization tests with environments different from those experienced during evolution. They have higher fitness during evolution and their capacity to make a detour around the obstacle and reach the target is more robust and generalizable.

However, it is not clear that the superior performance of the organisms with memory can be attributed to the fact that they store a trace of the position of the target and use this trace to guide their behavior when they do not have direct visual access to the target. The memory mechanism of our neural network is a cumulative one, which implies that the memory trace of each successive events is superimposed on the trace of the previous events. Therefore, the memory trace of an event becomes weaker and weaker as the event recedes back in time. Since our organisms can lose sight of the target for a long sequence of input/output cycles, the memory trace of the position of the target inevitably fades away and is unlikely to be used to guide the robot's behavior.

Our more detailed analysis of the behavior of our robots seems to indicate that the advantage of possessing a memory mechanism for solving the detour task lies elsewhere. This analysis has shown that the robots without memory tend to approach the ostacle and/or the peripheral wall more often than the robots with memory during the trajectory that brings them to the target. In fact, their proximity sensors are more activated than those of the robots with memory both during the early portion of the trajectory when the robots have to negotiate the obstacle and in the later portion when the robots can directly approach the target. The reliance on sensory input from the obstacle and the wall allows the robots without memory to solve the task but causes them to exhibit a less efficient performance than the robots with memory. They need to go near the obstacle or the wall to get access to sensory input from the obstacle or the wall to guide their behavior, and this inevitably slows down their performance because they take more time to reach the target.

The organisms with memory are in a different situation. Even if the external input from the environment does not change from one cycle to the next cycle when they have no visual access to the target and no sensory input from the obstacle or the perpheral wall, and therefore the external input cannot guide their behavior, these robots have access to an internal input from their memory units which docs change from cycle to cycle. Therefore, in the «blind» cycles (i.e., the cyles without external input), the robots with memory can use the varying internal input to generate the appropriate behavior. The result is that they do not need to waste time approaching the obstacle and/or the peripheral wall and they can generate shorter trajectories to the target. Hence, their superior performance both during evolution and in the generalization tests.

Returning to the notion of an «internal representation» as a necessary prerequisite for solving a detour task, we conclude that this notion is not necessary, at least in the conditions examined in our simulations. Simple feedforward neural networks can solve the task, although not very efficiently. But even the more efficient performance in the detour task of neural networks with memory units does not require a notion of internal representation. One might think that neural networks with memory units store an internal representation of the environment (more precisely, of the position of the target relatively to the organism) so that the network can use this internal representation when is has no direct visual access to the target. But our analyses have shown that the superior performance of the networks with memory units is probably determined not by a internal representation of the position of the target but by a simpler and more basic mechanism according to which networks with memory units can rely on the varying internal input from the memory units when the external input is not available - an option not open to networks without memory units.

References

1. Donnart, J. Y., Meyer, J. A. Learning reactive and planning rules in a motivationally autonomous animat. IEEE Transactions on Systems, Man, and Cybernetics-Part B 26(3) (1996) 381-395.
2. Dorigo, M., Colombetti, M. Robot shaping: Developing autonomous agents through learning. Artificial Intelligence 71(2) (1994) 321-370
3. Ellen, P., Thinus-Blanc, C. (eds.): Cognitive processes and spatial orientation in animal and man. Dordrecht, Martinus Nijkoff (1987)
4. Elman, J.L.: Finding structure in time. Cognitive Science 14 (1990) 179-211
5. Floreano, D., Mondada, F.: Evolution of homing navigation in a real mobile robot. IEEE Transactions on Systems, Man, and Cybernetics-Part B 26(3) (1996) 396-407.
6. Gallistel, C. R.: The Organization of Learning. Mit Press, Cambridge (1990)
7. Nolfi, S., Floreano, D., Miglino, O., Mondada, F.: How to evolve autonomuos robots: Different approaches in evolutionary robotics. In Brooks R. A., Maes P. (eds.): Artificial Life IV. Proceedings of the Fourth International Workshop on The Synthesis and Simulation of Living Systems. MIT press, Cambridge (1994) 190-197
8. Regolin, L., Vallortigara, G., Zanforlin M.: Object and Spatial representations in detour problems by chicks. Animal Behavior 48 (1994) 1-5
9. Thinus-Blanc, C.: Animal Spatial Cognition. World Scientific, Singapore (1996)
10. Tolman, E. C.: Cognitive maps in rats and men. Psychological Review 36 (1948) 13-24

Evolving Robot Behaviours with Diffusing Gas Networks

Phil Husbands

School of Cognitive and Computing Sciences
and
Centre for Computational Neuroscience and Robotics
University of Sussex, Brighton, UK
philh@cogs.susx.ac.uk

Abstract. This paper introduces a new type of artificial nervous system and shows that it is possible to use evolutionary computing techniques to find robot controllers based on them. The controllers are built from networks inspired by the modulatory effects of freely diffusing gases, especially nitric oxide, in real neuronal networks. Using Jakobi's radical minimal simulations, successful behaviours have been consistently evolved in far fewer evaluations than were needed when using more conventional connectionist style networks. Indeed the reduction is by a factor of roughly one order of magnitude.

1 Introduction

In evolutionary robotics, the predominant class of systems for generating behaviours is that of artificial neural networks (ANNs). These networks can be envisaged as simple nodes connected together by directional wires along which signals flow. The nodes perform an input output mapping that is usually some sort of sigmoid function [7]. Occasionally a simple differential equation is used instead, providing the possibility of richer dynamics [1]. Some have used feedforward architectures, others have explored more free form arbitrarily recurrent networks. The original inspiration for all these styles of network is the neuroscience of the 1940s and 1950s. They abstract something of the electrical properties and behaviours of real neuronal networks. However, two obvious questions raise their heads. Do we have any reason to believe that these kinds of systems are capable of generating adaptive behaviours in autonomous robots of a kind that is much more sophisticated than we can manage today? Even if they are, will it be possible to find the networks in question, through evolutionary search or some other technique?

As has been pointed out by various people (e.g. [3]), advances in neuroscience have made it clear that the propagation of action potentials, and the changing of synaptic connection strengths, is only a very small part of the story of the brain (e.g [16]). This in turn means that connectionist style networks, and even recurrent dynamical ones, are generally very different kinds of systems from those that generate sophisticated adaptive behaviours in animals. Indeed, it may be better to think of natural nervous systems as some kind of ever changing chemical machine [16]. Although our picture of biological neural networks changes every few years, advances over the past decade or so can provide a rich source of inspiration in devising alternative styles of artificial network. Among others, Brooks and colleagues went some way down that path by using ideas gleaned from some of the properties of the lobster hormonal system [2].

As far as the author of this paper is concerned, current understandings of nervous systems seem to suggest that a useful abstract way to think of them is in terms of several interacting classes of dynamical processes, each with distinct characteristics (e.g. electrical, short-range chemical, long-range chemical)[1]. These processes can have very different spatial and temporal properties and may be heavily intertwined, modulating each other in complex ways. For instance, Hebbian style changes in synaptic efficacy can be thought of as one of these processes that can be turned on and off and localised by other processes, such as the change in local concentration of a particular chemical. At the detailed level, there will be complex cascades of chemical reactions involved, but it is suggested that it may be possible to pull out more abstract systems that describe the gross dynamical principles underlying the behaviour of nervous systems. It is hoped that versions of these abstract systems, based on interacting dynamical processes, can be developed that are sufficient to underpin adaptive machines more advanced than those available today. It is very difficult to see how such systems can be investigated and developed except through some form of artificial evolution. It is likely that, at least in principle, for any of these systems a functionally equivalent recurrent dynamical network exists. However, my bet is that in general it will be horrendously convoluted and almost impossible to find. It is hoped that the kinds of systems just sketched will be more amenable to evolution than standard ANNs and will be far easier to scale up. We shall see.

This paper describes some very initial investigations into the kind of systems outlined in the previous paragraph. The main inspiration has been the recent discovery that freely diffusing nitric oxide (NO) is synthesised in, and emitted by, nerve cells in many parts of nervous systems. As it diffuses, it acts as a neurotransmitter and is implicated in the modulation of various properties of nerve cells and synapses [4].

After describing the networks inspired by this biological phenomenon, so called GasNets, a number of successful evolutionary robotics experiments using this new style of network are discussed. The paper closes with some conclusions, including the observation that, using GasNets, successful behaviours have been consistently evolved in far fewer evaluations than were needed when using more conventional connectionist style networks.

---

[1] ANNS incorporating Hebbian connection strength changes can be thought of as a simple example of this kind of system. I am proposing exploring far richer systems.

2   GasNets

In this section a new form of artificial nervous system inspired by two aspects of biological neuronal networks is described in detail. These two aspects are: the heterogeneity of intrinsic nerve cell properties found in much of the nervous system of invertebrates; and the modulation of these properties by diffusing NO, emitted from within the nervous system itself. Many of the known and postulated effects of NO are not included, but will be the inspiration for future developments of this style of network.

These are abstract systems founded on some of the general principles, rather than details, of biological networks. They are not models of real nervous systems. The abstractions chosen for this initial investigation into this style of network are: heterogeneity in terms of transfer functions at the nodes in the network, modulation of intrinsic properties in terms of changing these functions as the network runs. One last general point – these networks should not be thought of as computational devices. They could be thought of as mathematical systems, but I prefer to regard them much more as simulations of physical devices.

2.1   The 4 gas GasNet

The networks used in the experiments described later consist of units connected together by excitatory links, with a weight of +1, and inhibitory links, with a weight of -1. The output , O_j, of a node j is a function of the normalised sum of its inputs, S_j, as described by equation 1. In addition to this underlying network in which positive and negative 'signals' flow between units, an abstract process loosely analogous to the diffusion of gaseous modulators is at play. Some units can emit 'gases' which diffuse and are capable of modulating the behaviour of other units by changing their transfer functions in ways described in detail later. This form of modulation allows a kind of plasticity in the network in which the intrinsic properties of units are changing as the network operates. The networks function in a 2D plane; their geometric layout is a crucial element in the way in which the 'gases' diffuse and affect the properties of network nodes. This aspect of the networks is described in more detail later.

$$O_j = f(\overline{S_j}) \tag{1}$$

Where,

$$\overline{S_j} = \frac{\left(\sum_{p \in P_j} O_p - \sum_{n \in N_j} O_n + \sum_{k \in SEN_j} I_k\right)}{(np_j + nn_j + ns_j)} + R \tag{2}$$

In equation 2, P_j is the set of network elements with excitatory connections to element j. Likewise, N_j is the set of elements with inhibitory link to j, and SEN_j is the set of sensors connected to j. R is the default activation of a node (= 0.05). np_j, nn_j and ns_j are , respectively, the number of positive, negative and sensor connections to element j, i.e.:

$$np_j = |P_j| \tag{3}$$

$$nn_j = |N_j| \tag{4}$$

$$ns_j = |SEN_j| \tag{5}$$

Normalizing by dividing by the number of inputs keeps the summed input in the range [-1,1]. The transfer function, f, is defined in equation 6, its output range is [-1,1] given the restriction on the input range.

$$f(x) = \begin{cases} 0, & \text{if } x < 0 \text{ and } (a < 0 \text{ or } b < 0) \\ (x^a + x^b)/2, & \text{otherwise} \end{cases} \tag{6}$$

Where,

$$a, b \in PP = \{0.1, 0.2, 0.3 \dots 0.8, 1, 2, 3 \dots 9, 10\} \tag{7}$$

Overlays of plots of this function for many of the possible combinations of a and b are shown in figure 1. As can be seen, a wide range of output responses to a given input are possible, depending on the values of the parameters a and b. As will seen later, default values of a and b for each node are set genetically, but are changed by diffusing gases as the network runs. It is genetically determined whether or not a node will emit one of four gases, and under what circumstances emission will occur (either when the 'electrical' activation of the node exceeds a threshold, or the concentration of one of the gases, genetically determined, in the vicinity of the node exceeds a threshold).

Gas Diffusion in the Networks A very abstract model of gas diffusion is used. For an emitting node, the concentration of gas at distance d from the node is given by equation 8. Here r is the genetically determined radius of influence of the node, so that concentration falls to zero for $d > r$. $TC(t)$ is a linear function that models build up and decay of concentration after the node has started/stopped emitting. The slope of this function is individually genetically determined for each emitting node, C_0 is a global constant.

$$C(d, t) = \begin{cases} C_0 \times e^{\frac{-2d}{r}} \times TC(t), & d < r \\ 0, & \text{otherwise} \end{cases} \tag{8}$$

$$TC(t) = \begin{cases} H(\frac{(t - t_e)}{k}), & \text{emitting} \\ H(H(\frac{(t_s - t_e)}{k}) - H(\frac{(t - t_s)}{k})), & \text{not emitting} \end{cases} \tag{9}$$

Where, t_e is the time at which emission was last turned on, t_s is the time at which emission was last turned off, k is genetically determined for each node and:

$$H(x) = \begin{cases} x, & x < 1 \\ 0, & x \le 0 \\ 1, & \text{otherwise.} \end{cases} \tag{10}$$

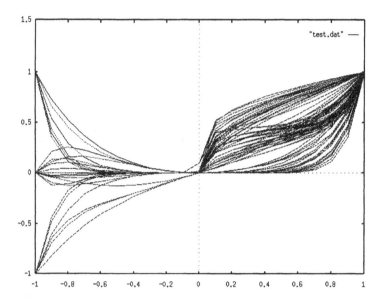

Fig. 1. The transfer function used for the 4 gas GasNets. Input on the X axis, output on the Y. Overlays of the function (see equation 6) for many combinations of a and b are shown.

In other words, the 'gas' concentration varies spatially as a Gaussian centred on the emitting node. The height of the Gaussian at any point within the circle of influence of the node is linearly increased or decreased depending on whether the node is emitting or not. Note $TC(t)$ saturates at a maximum of 1 and a minimum of 0.

Modulation by the Gases The values of a and b in the network unit transfer function (see equation 6) are changed (or *modulated*) by the presence of gases at the site of a unit. This modulation is described by equations 12–17 and happens continually as the network runs. This provides a form of plasticity very different from that found in more traditional artificial neural networks. At every time step the value of a for node$_i$, a^i, is updated according to equation 11.

$$a^i = PP[a^i_{index}]$$ (11)

Where,

$$a^i_{index} = S(N, a^i_{ni})$$ (12)

Here a^i_{index} is node$_i$'s index into the set PP shown in equation 7. N is the number of elements in PP. At each time step a^i_{ni} is updated according to equation 13. The linear (thresholded) function S is described by equation 14.

$$a_{ni}^i = a_{def.index}^i + \frac{C_1}{C_0 \times K} \times (N - a_{def.index}^i) - \frac{C_2}{C_0 \times K} \times (a_{def.index}^i) \quad (13)$$

Where $a_{def.index}^i$ is the genetically set default value for a_{index}^i, C_1 is the concentration of gas 1 at the site of node$_i$, C_2 is the concentration of gas 2 at the site of node$_i$ and C_0 and K are global constants. So, a_{index}^i increases in direct proportion to the concentration of gas 1, and decreases linearly with respect to the concentration of gas 2. In this way the value of a for node$_i$ is changed by the presence of gases 1 and 2 at the node's site.

$$S(N, x) = \begin{cases} x, & 0 \leq x \leq N \\ 0, & x < 0 \\ N, & x > N \end{cases} \quad (14)$$

Similarly the value of b at node$_i$, b^i, is changed by the presence of gases 3 and 4. This is described by equations 15–17.

$$b^i = PP[b_{index}^i] \quad (15)$$

$$b_{index}^i = S(N, b_{ni}^i) \quad (16)$$

$$b_{ni}^i = b_{def.index}^i + \frac{C_3}{C_0 \times K} \times (N - b_{def.index}^i) - \frac{C_4}{C_0 \times K} \times (b_{def.index}^i) \quad (17)$$

2.2 The 2 gas GasNet

This paper will concentrate on early results found using the 4 gas style GasNet. However, we are currently investigating slightly simpler versions using 1 or 2 gases. For instance, in one of our 2 gas networks the transfer function at each node is of the form shown in equation 18[2]. Again the weights are restricted to be either +1 or -1. The value of b is genetically set at each node as is the default value of k. The two gases raise and lower the value of k in a similar manner to the way a and b are changed in the 4 gas model.

$$O_j = tanh([k \times \sum_i O_i w_{ij}] + b) \quad (18)$$

---

[2] Thanks to Andy Philippides for suggesting this form.

3   Network encoding

The basic geneotype used to encode 4 gas GasNets consisted of an array of real numbers. Each node in the network had 15 real values associated with it. That is:

$< geneotype > :: (< gene >)^*$

$< gene > :: < x >< y >< R_p >< \Theta_{1p} >< \Theta_{2p} >< R_n >< \Theta_{1n} >< \Theta_{2n} >$
$< rec >< TE >< CE >< k >< R_e >< a_{def.ind} >< b_{def.ind} >$

This encoding was used to generate networks conceptualized to exist on a 2D Euclidean plane. x and y give the position of a network node on the plane. The next six numbers define two segments of circles, centred on the node. These segments are used to determine the connectivity of the network. R_p gives the radius of the 'positive' segment, Θ_{1p} its angular extent and Θ_{2p} its orientation. R_n, Θ_{1n} and Θ_{2n} define a 'negative' segment. The radii range from zero to half the plane dimension, the angles range from zero to 2π. The segments are illustrated in figure 2. Any node that falls within a positive segment has an excitatory (+1) link made to it from the segment's parent node. Any node that falls within a negative segment has an inhibitory (-1) link made to it from the segment's parent node. Developing networks on a 2D plane has a number of advantages as discussed in [11]. The idea of using a plane has been previously explored by at least Husbands [8], Nolfi [15] and Jakobi [10]. Of course, in this case the geometry of the network is crucial to the workings of the gas diffusion, so the networks *have* to exist on a plane. This particular encoding is directly inspired by Jakobi's [11].

The rest of a gene is interpreted as follows. The value of $rec \in \{0,1,2\}$ determines whether the node has no recurrent connection to itself, an excitatory recurrent connection or an inhibitory recurrent connection, respectively. $TE \in \{-1,0,1,2,3,4\}$ provides the circumstances under which the node will emit a gas. These are: not at all, if its 'electrical' activity exceeds a threshold, or if the concentration of the referenced gas (1,2,3 or 4) at the node site exceeds a threshold. $CE \in \{1,2,3,4\}$ gives the gas the node can emit. k is a real in the range [1,15] and is used to control the rate of gas build up/decay as described earlier by equation 9. R_e is the maximum radius of gas emission, this ranges from 2 to half the plane dimension. $a_{def.ind}$ and $b_{def.ind}$ are the default values for the a and b indices as used in equation 11 to determine the default values of a and b for the node.

This basic encoding can be used to search for network topologies and geometries with a fixed number of nodes, or a dynamic length version can be used to evolve networks in a more open-ended way [6]. Of course, the use of the segments to determine the connectivity means the number of connections, and hence the basic architecture, is never fixed. To say nothing of the additional dynamic properties introduced by the diffusing gases.

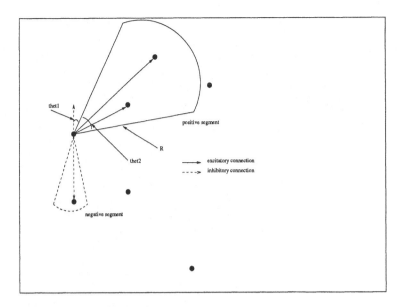

Fig. 2. Positive and negative segments define the connectivity of the network. The network develops and functions on a 2D plane.

4 Experiments

In order to start exploring the properties of GasNets, particularly their evolvability and suitability as control systems, it was decided to rerun some of Jakobi's recent experiments [10, 11], substituting the new type of network for his more conventional connectionist ones. It would then be possible to compare the kinds of solutions found, how quickly they evolved, and so on. A major reason for choosing Jakobi's work was the fact that he had evolved the behaviours in question using his innovative and radical minimal simulations [11, 10, 12]. Control systems evolved using these ultra-lean ultra-fast simulations transfer perfectly to reality. Because of their speed, many evolutionary runs can be performed allowing the kind of exploration desired in this case. Jakobi's original minimal simulation code was used for the experiments described here.

4.1 Khepera with state: The T-maze

The first behaviour attempted was that required to perform Jakobi's T-maze task. This is illustrated in figure 3. A Khepera robot [13], making use of 6 IR proximity sensors and 2 ambient light sensors, moves along a corridor. A light shines from either the left or the right, chosen randomly. Once the Khepera reaches the T-junction it should turn in the direction the light shone from.

For this experiment, following Jakobi [10], a fixed number of network nodes was used. Also, bilateral symmetry was imposed on the control network. 14 nodes

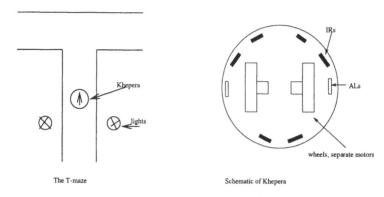

The T-maze   Schematic of Khepera

Fig. 3. Jakobi's T-maze task. At the junction the robot must turn in whichever direction the light was shining from. Schematic of Khepera, showing sensor positions, at the right of the figure.

were used. One for each of the sensors, 2 for each motor (motor output was calculated as forward-motor-node output - backward-motor-node output), and 2 others. Bilateral symmetry was achieved by encoding for 7 nodes. Their (x,y) coordinates were constrained to lie in the left half of the network plane. Each of the nodes was reflected in the line x=(half plane width), creating 7 new nodes. The positive and negative segments of the original nodes were also reflected in the half way line. All other genetically encoded properties were inherited intact. On developing the connectivity of the network, as determined by the segments, a symmetrical network is formed. The left hand side is connected to the Khepera's left hand sensors and motor, the right hand side is connected to the Khepera's right hand sensors and motor. One predetermined node for each sensor and motor.

A distributed, or diffusion, GA of the kind described in [9] was used. It employed a population of 225 spread out over a 15x15 grid with overlapping Gaussian local neighbourhoods, in which local selection rules operated. Standard one point crossover was used with a probability of 0.9. Mutation operated as follows. With a rate of one mutation per geneotype, any value to be mutated was changed by a random real in the range $\pm 10\%$ of its total range 80% of the time and by a random value in the range $\pm 40\%$ the remaining 20% of the time.

Each evaluation consisted of 12 trials. The starting orientation of the robot, the corridor width and length and the position of the lights were all randomly varied between trials. Other aspects of the simulation were randomly varied within and across trials in keeping with the minimal simulation methodology [10]. The fitness function (taken from [10]) is shown in equation 19.

$$f = d_1 + d_2 + bonus \qquad (19)$$

Where d_1 is the distance moved down the first corridor, d_2 is the distance moved down the second corridor and *bonus* is 100 if the robot turns in the correct direction at the junction, 0 if it doesn't. A trial is aborted if the robot touches a wall.

Jakobi had originally used binary networks in which the connectivity, weights on the connections and node thresholds were genetically encoded [10]. He had been able consistently evolve robust successful controllers in 1,000 generations using a similar GA to that described above, with a population of 100. This may seem rather a large number of evaluations, but it must be appreciated that evolving in a minimal simulation i sin many senses harder than in reality, because of the extreme use of noise. Hence it would be expected that a greater number of evaluations would be required than in reality, although this is heavily offset by the speed at which the simulations run. The resulting controllers are extremely robust, capable of successful behaviour in a wide range of conditions.

To date more than 15 runs have been completed using the GasNets. In each run success was achieved in less than 100 psuedo generations[3], in several cases in less than 50. *In other words, the number of evaluations needed was decreased by a factor of one order of magnitude.*

Interestingly, some of the successful controllers made use of the gases and some didn't. All were very heterogeneous as far as the transfer functions of the nodes were concerned. In the runs where the final successful controllers didn't use gases, they *had* been used extensively during the evolution of the final population. When the gases were removed all together, it was not possible to evolve successful controllers within 500 generations. Since the gases are the agents of a form of plasticity in the networks, these observations suggest that the Baldwin effect [14] is at play, albeit a rather different form than has been observed before. This will need further investigation.

Many different successful behaviour generating mechanisms were observed. The evolved controller shown in figure 4 represents a class of mechanisms, making use of the gases, of which several examples were seen.

Briefly, the network shown in figure 4 works as follows. The motor transfer functions are such that the default behaviour, in the absence of sensory input, is straight line forward motion. The left most and right most IR sensors (see figure 3) are connected directly to the right and left motors, respectively. This arrangement, coupled with the particular motor transfer functions, provides a basic Braitenburg style obstacle avoidance behaviour. This is what enables the robot to travel along the corridors without crashing. The brief chemical activity initiated at nodes 1 and 8 by the IR stimulus, and the ensuing transfer function modulation, reinforces the Braitenburg behaviour. Nodes 3 and 10 are connected to the left and right ambient light sensors respectively. When either is stimulated by the light shining from the side of the first corridor it feeds into the nodes 0 and 7 feedback loop. The transfer functions are such that only a continuous stimuli ramps up the activity levels and initiates the release of gas 2 from whichever of nodes 0 and 7 is being directly fed the active AL signal. The evolved time constants on the gas build up and decay processes mean that only the continuous AL stimuli as the robot moves through the light zone (as opposed

---

[3] One psuedo generation occurs every N offspring events, where N = population size.

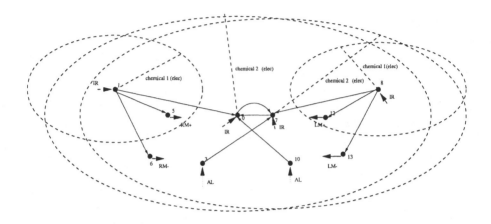

Fig. 4. An evolved successful control network for the T-maze task. Network connectivity plus the radius of influence of emitting nodes is shown. These are depicted as ellipses as the X scale is twice the Y scale. See text for details.

to the very brief IR stimuli from nodes 1 and 8 ensured by the successful obstacle avoidance subnetwork) triggers this part of the circuit. If the left AL sensor is active it triggers release of gas 2 from node 7. Because node 7 is closer to the left motor nodes (12 and 13) than the right motor nodes (5 and 6) it has a stronger modulatory effect on their transfer functions. The overall effect is to reduce left motor output with respect to right. By the time the robot comes to the junction it inevitably turns to the left. In a similar way, if the right AL sensor had been on, the robot would have turned to the right. The underlying Braitenburgh obstavcle avoider still works under this modulation. The genetically determined slow decay of gas 2 is important to the successful operation of this behaviour. Obviously the geometric layout of the network is crucial. Unfortunately space does not allow a more detailed analysis here. Figure 5 shows the transfer functions for the network. Note that they cover a very wide range of response types. This was found in every successful controller examined, including those that did not use gases. This, along with the observed role that the differences in transfer functions plays in successful controllers, suggests transfer function heterogeneity can be a very useful thing in sensorimotor systems. All of this is moving some way from the kind of positions held in mainstream cognitive science as to what a behaviour generating mechanism should look like. It adds more grist to the mills of left-field philosophers of cognitive science such as Wheeler [17].

4.2 Khepera with more state, the double T-maze

The second behaviour attempted was that required to perform the double T-maze task. This is illustrated in figure 6.

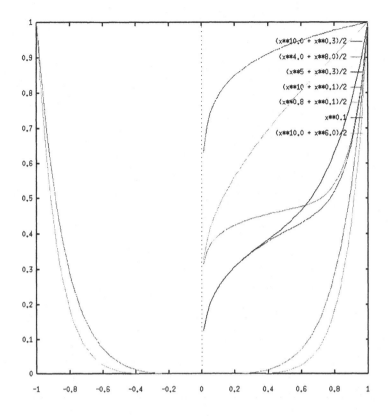

Fig. 5. The transfer functions at the nodes of the controller shown in figure 4

This time there is a sequence of 2 lights. The robot must turn at a pair of junctions in accordance with the directions from which the two lights are shining (e.g. right,left). The experimental setup was essentially the same as for the single T-maze, with a slightly modified fitness function.

$$f = d_1 + d_2 + d_3 + bonus_1 + bonus_2 \qquad (20)$$

Jakobi was not able to evolve successful controllers for this task using his binary networks. However, with the GasNets, for 5 runs completed to date, success was achieved in 3 runs by 350 generations; in 1 run success was achieved by 700 generations; and in 1 run success was not achieved by 1000 generations. At the time of writing none of these runs has yet been analysed.

4.3 Visually guided behaviours

In this section a very brief mention of some on-going joint work with Tom Smith will be made. We are in the process of investigating the use of GasNets to control

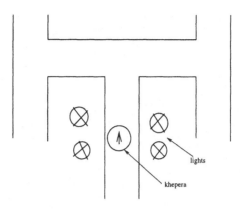

Fig. 6. The double T-maze.

a visually guided robot. The Sussex gantry robot [5] is best thought of as a two wheeled device with a fixed forward pointing video camera. We have been concurrently evolving network controllers and the robot's visual morphology (the genetically specified number and positions of pixels from the camera's image that provide the only sensory input to the robot). We are using another of Jakobi's minimal simulations [10] to evolve a target discriminating behaviour (move to a triangle while ignoring a rectangle) under very noisy lighting conditions.

Jakobi reported needing 6,000 generations of his GA to reliably evolve robust controllers. We have found successful controllers in less than 800 generations using the 4 gas GasNets and the same GA and encoding described earlier. This time the GA was allowed to find the appropriate number of network nodes by using gene insertion and deletion operators that allowed the geneotype length to vary. A successful controller (complete with evolved visual morphology) is shown in figure 7. It is structurally much less complex than Jakobi's evolved controllers. Indeed it seems remarkably simple considering the very noisy nature of the lighting and the relative complexity of the task. This work will be reported on in detail elsewhere and is mentioned here to demonstrate that GasNets are not merely restricted to good performance on one type of task.

5   Conclusions and Discussion

This paper has introduced a new type of artificial nervous system and has shown that it is possible to use evolutionary computing techniques to find robot controllers based on these systems. This has been demonstrated for a range of behaviours involving two different robots. These successes were achieved using very similar setups to those employed by Jakobi when he previously evolved controllers for the same tasks. The main (although admittedly not only) difference between our experiments and his were the style of network used. He used

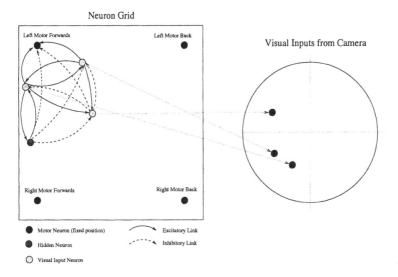

Fig. 7. Successful evolved GasNet for a visually guided behaviour. See text for details.

fairly standard connectionist type networks while we used GasNets, as described earlier in this paper. We found that we were able to consistently evolve success-ful controllers in far fewer evaluations than him. Indeed the reduction was by a factor of roughly an order of magnitude.

This suggests that the space of possible behaviours open to being generated by GasNets is somehow 'thicker' than for Jakobi's more conventional networks. It is easier to find successful controllers in this space; it is rich with useful network dynamics and mechanisms.

This in turn suggests that networks involving a number of interacting dy-namical processes with distinct properties and characteristics, may well be a very powerful building block for evolutionary robotics. The inspiration for these networks were, of course, real nervous systems. Recent advances in neuroscience mean that there is a rich seam of inspiration to mine. In abstracting principles from biological systems in order to build adaptive machines, it is hoped that a mutually beneficial interface between neuroscience and AI will flourish and prove to be profoundly important in both fields. The initial studies reported here are the very beginning of a new line of research at Sussex. There is much more of this story to come.

Acknowledgements

Many thanks to Nick Jakobi for his minimal simulation code. Without his in-novative techniques this study would not have been possible. This exploratory study would not have been possible without it. Many thanks to Mick O'Shea

(Co-Director with me of the Sussex CCNR) and other colleagues from the School of Biological Sciences for introducing me to some of the more exotic avenues of contempary neuroscience, especially the role of NO. Thanks to Tom Smith for figure 7 and for collaboration on the studies very briefly mentioned in section 4.3. Most of this work was done while I was at UniVap, Sao Jose dos Campos, SP, Brazil. I gratefully acknowledge the sponsorship of the British Council (SPA/126/881) and of FAPESP (proj. 96/7200-8) and pay tribute to Pedro Paulo Balbi de Oliveira's unstinting hospitality.

References

1. R.D. Beer and J.C. Gallagher. Evolving dynamic neural networks for adaptive behavior. *Adaptive Behavior*, 1(1):91–122, 1992.
2. R. A. Brooks. Challenges for complete creature architectures. In J.-A. Meyer and S.W. Wilson, editors, *From Animals to Animats: Proceedings of The First International Conference on Simulation of Adaptive Behavior*, pages 434–433. MIT Press/Bradford Books, Cambridge, MA, 1991.
3. R. A. Brooks. Coherent behavior from many adaptive processes. In D. Cliff, P. Husbands, J.-A. Meyer, and S.W. Wilson, editors, *From Animals to Animats 3: Proceedings of The Third International Conference on Simulation of Adaptive Behavior*, pages 22–29. MIT Press/Bradford Books, Cambridge, MA, 1994.
4. J. Garthwaite. Glutamate, nitric oxide and cell-cell signalling in the nervous system. *Trends in Neuroscience*, 14:60–67, 1991.
5. I. Harvey, P. Husbands, and D. Cliff. Seeing the light: Artificial evolution, real vision. In D. Cliff, P. Husbands, J.-A. Meyer, and S. Wilson, editors, *From Animals to Animats 3, Proc. of 3rd Intl. Conf. on Simulation of Adaptive Behavior, SAB'94*, pages 392–401. MIT Press/Bradford Books, 1994.
6. Inman Harvey. Evolutionary robotics and SAGA: the case for hill crawling and tournament selection. In C. Langton, editor, *Artificial Life III*, pages 299–326. Santa Fe Institute Studies in the Sciences of Complexity, Proceedings Vol. XVI, Addison-Wesley, Redwood City CA, 1994.
7. G.E. Hinton, J.L. McClelland, and D.E. Rumelhart. Distributed representations. In D.E. Rumelhart, J.L. McClelland, and the PDP Research Group, editors, *Parallel Distributed Processing: Explorations in the Microstructure of Cognition*, pages 77–109. MIT Press, 1986.
8. P. Husbands, I. Harvey, D. Cliff, and G. Miller. The use of genetic algorithms for the development of sensorimotor control systems. In P. Gaussier and J-D. Nicoud, editors, *Proceedings of From Perception to Action Conference*, pages 110–121. IEEE Computer Society Press, 1994.
9. Philip Husbands. Distributed coevolutionary genetic algorithms for multi-criteria and multi-constraint optimisation. In T. Fogarty, editor, *Evolutionary Computing, AISB Workshop Selected Papers*, pages 150–165. Springer-Verlag, Lecture Notes in Computer Science Vol. 865, 1994.
10. N. Jakobi. Evolutionary robotics and the radical envelope of noise hypothesis. *Adaptive Behavior*, 6(2):(in press), 1998.
11. N. Jakobi. *Minimal Simulations for Evolutionary Robotics*. PhD thesis, University of Sussex, 1998.

12. N. Jakobi. Running across the reality gap: Octopod locomotion evolved in a minimal simulation. In P. Husbands and J.-A. Meyer, editors, *EvoRobot98*, page (this vol.). Springer-Verlag, 1998.

13. K-Team. Khepera users manual. EPFL,Lausanne, June 1993.

14. G. Mayley. Landscapes, learning costs and genetic assimilation. *Evolutionary Computation*, 4(3), 1996.

15. S. Nolfi D. Parisi. Evolving artificial neural networks that develop in time. In F. Moran, A. Moreno, and J.J. Merelo, editors, *Advances in Artificial Life: Proceedings of the third European conferrence on Artificial Life.*, Berlin, 1995. Springer-Verlag.

16. D. Purves and George Augustine. *Neuroscience.* Sinauer, 1997.

17. M. Wheeler. From robots to rothko: The bringing forth of worlds. In M. Boden, editor, *The Philosophy of Artificial Life*, pages 209–236. OUP, Oxford, 1996.

Explaining the Evolved: Homunculi, Modules, and Internal Representation

Michael Wheeler

Department of Experimental Psychology
University of Oxford
South Parks Road, Oxford, OX1 3UD, UK
Phone: +44 1865 271386
Fax: +44 1865 310447
E-Mail: michaelw@psy.ox.ac.uk

1 Introduction

Evolutionary robotics is the discipline in which algorithms inspired largely by Darwinian evolution are used to automatically design the control systems for artificial autonomous agents.[1] The field is often portrayed as an engineering endeavour. Understood as such, the ultimate goal of the work is to produce robots that do useful things. But although this engineering mission is undoubtedly a large part of the discipline's intellectual profile, evolutionary robotics has an alternative identity, as a research programme within which certain explanatory ideas, about how adaptive-behaviour-generating control systems might work, are developed and tested. Given that biological nervous systems can be treated as a sub-class of such control systems, these explanatory ambitions promise to have important consequences for cognitive science. In what follows, I shall be concerned primarily with evolutionary robotics in this alternative guise (although, as we shall see, the explanatory ambitions of the field are deeply intertwined with its engineering expectations). My intention is to investigate, from a philosophical perspective, a particular argument, one that draws on evolutionary robotics in order to cast doubt on the idea that the concept of *internal representation* will be a key term in any proper scientific understanding of adaptive behaviour. Before I lay out this target argument, a few preliminary remarks about representations are in order.

Agents often represent their worlds. More informatively, agents often represent their worlds, in the sense that they regularly *take their worlds to be a certain way*. Thus a person may represent as disgraceful the hunting of animals for sport, or a prey may represent as a threat the presence of a predator. This sort of representation-talk, pitched, as it is, at the level of whole agents, appears to be uncontroversial and indeed useful, since sometimes it helps us to understand why people and non-human animals behave the way they do. In cognitive science, however, it is not this mundane, primary sense of representation that

---

[1] For those who are new to evolutionary robotics, two useful points of entry are [10] and [21].

bears the explanatory burden, but a second sense of the term, according to which representations are conceptualised as internal states of a behaviour-controlling system, states whose functional role is to *stand in for* (usually external) objects and situations (such as predators and threats) in that system's internal goings-on. The idea is that the presence of such *internal vehicles of representational content* (as philosophers are inclined to call such states) explains, in a wholly scientific manner, the primary phenomenon of cognitive representation (i.e., how it is that the agent takes its world to be a certain way). This second sense of representation — to which the term *internal representation* shall, in this paper, refer — has, famously, come under fire in recent years, from sources as diverse as developmental psychology (e.g., [28]), philosophy of cognitive science (e.g., [18, 34]), behaviour-based robotics (e.g., [33]), and, most importantly of all, given the present context, *evolutionary robotics* (e.g., [1, 2, 12, 15]).

The specific anti-representational argument that will concern us here can be stated in the following, general form:

1. The most theoretically compelling account of internal representation tells us that internal representations are present only in those adaptive-behaviour-generating control systems (henceforth just *control systems*) that succumb to a particular style of decomposition (to be identified later).
2. Given evolutionary building blocks of a certain kind (also identified later), artificial evolution will, in general, design control systems which will be stubbornly resistant to this style of decomposition.
3. So the concept of internal representation will, in general, not be useful for analysing control systems that have been produced by artificial evolution working with that kind of evolutionary building block.
4. There are enough abstract similarities between (a) the class of evolved artificial control systems in question and (b) the corresponding class of evolved biological control systems (in terms of, for instance, the nature of the evolutionary building blocks used in each case), to mean that conclusions reached by studying the former will apply equally to the latter.
5. So, in general, the concept of internal representation will not be useful for analysing the control systems produced by natural evolution, i.e., biological nervous systems.[2]

I am going to take this argument and give it a routine examination, just to see how it is getting on. It is not the aim of the exercise to make any final pronouncements on how things will turn out in the end, since, in the nascent area of research that is evolutionary robotics, any attempt to do so would be ridiculously premature and, therefore, unhelpful. Rather the aim is to clarify some of the issues and questions that matter. The region to be explored is one

---

[2] As far as I am aware, the first person to present this sort of position in any detail — although never in quite the form used here — was Inman Harvey (see [15, 16]). I defended the view in [34]. This paper returns to the issues broached there, with the aim of modifying and/or adding to the arguments, as now appears necessary or appropriate, given the intervening years of philosophical and scientific research.

that many readers will, at first sight anyway, judge to be painfully familiar. But even though a large amount of philosophical and scientific time and effort may already have been spent on trying to assess the status of internal representations, the vaguely deplorable fact is that there are deep and difficult problems hereabouts that remain unresolved. We have a duty to revisit those problems. Let's begin, then, with the issue of just what sort of control system might be ripe for representational pickings.[3]

2 Inner Voices

There is a style of explanation that (following others, and for the want of a better term) I shall call *homuncular explanation* (or *homuncular decomposition* or *the homuncular strategy*). This way of thinking is part of the conceptual profile of almost all research in cognitive science. In the case of classical (i.e., physical-symbol-system) cognitive science, it is simply written into the rule book. In the case of connectionist cognitive science, homuncular explanation is not mandatory. The straightforward fact, however, is that the vast majority of connectionists retain a commitment to the idea (see below).

For my money, the classic statement of homuncular decomposition is the following little vignette from Fodor.

> This is the way we tie our shoes: There is a little man who lives in one's head. The little man keeps a library. When one acts upon the intention to tie one's shoes, the little man fetches down a volume entitled *Tying One's Shoes*. The volume says such things as: "Take the left free end of the shoelace in the left hand. Cross the left end of the shoelace over the right free end of the shoelace ...," etc. ... When the little man reads "take the left free end of the shoelace in the left hand," we imagine him ringing up the shop foreman in charge of grasping shoelaces. The shop foreman goes about supervising that activity in a way that is, in essence, a microcosm of tying one's shoe. Indeed, the shop foreman might be imagined to superintend a detail of wage slaves, whose functions include: searching representations of visual inputs for traces of shoelace, dispatching orders to flex and contract fingers on the left hand, etc.. [14, pp.23-4].

According to this explanatory strategy, then, if we as cognitive scientists wish to understand how a whole agent performs a complex task (e.g., finding food, avoiding predators, tying shoes), we should proceed by analysing that complex

---

[3] It will, in fact, be some time before we turn our attention explicitly to evolutionary robotics. I apologise in advance to those readers who will grow impatient at the wait, but the lengthy set-up is, it seems to me, necessary, since one can't hope to evaluate the implications of evolutionary robotics for representationalism in cognitive science, without first getting a firm intellectual grip on how that sort of representationalism works.

task into a number of simpler sub-tasks (each of which has a well-defined input-output profile), and by supposing that each of these sub-tasks is performed by an internal 'agent' less sophisticated than the actual agent. These internal 'agents' are conceptualised as communicating with each other, and thus as co-ordinating their collective activity so as to perform the overall task. This first-level decomposition is then itself subjected to homuncular analysis. The first-level internal 'agents' are analysed into committees of even simpler 'agents', and each of these 'agents' is given an even simpler task to perform. This progressive simplification of function continues at increasingly lower levels until, finally, the sort of thing which you are asking each of your 'agents' to do is something so primitive that the explanation is almost certainly going to be neurobiological rather than psychological.

Philosophically speaking, this 'bottoming-out' in neurobiology is important. It is supposed to prevent the homuncular model from committing itself to the debacle of an infinite regress of systems, each of which, in order to do what is being asked of it, must literally possess the very sorts of intentional capabilities (e.g., the capacity to understand the meanings of messages) that the model is supposed to explain (see, e.g., [13]). In other words, the 'bottoming-out' is supposed to ensure that all talk of 'little people in the head' remains entirely metaphorical. This is an issue to which we shall return. For now let's press on ahead, and strip away the metaphorical veneer, in order to reveal the core elements of the homuncular strategy. Faced with the problem of explaining how an agent solves a complex task, the cognitive scientist engaging in homuncular explanation compartmentalises that agent into specialised subsystems with the following properties: (i) they are *well-defined* (in that they solve particular, well-defined sub-tasks, and thus have, what one might call, well-defined functional boundaries); (ii) they *communicate* with each other; and (iii) they are *hierarchically organised.*

3 Representations: When Modularity Becomes Homuncularity

It is time now to pin down, as best we can, the relationship that exists between homuncular decomposition and internal representation. It would be an understatement to say that this issue is still a matter of debate (see, e.g., [1, 8, 9, 15, 31, 34]), but it seems (to some of us at least) that the relation in question has turned out to be one of mutual support. To see why this is, one has to pick a very careful path through some slippery arguments and distinctions. And that's what I shall attempt to do, in this section.

Once upon a time I was convinced that internal representation ought to be part of the *definition* of homuncular decomposition. Thus I characterised homuncular decomposition as the strategy in which one explained the behaviour of a complex system by compartmentalising that system into a hierarchy of specialised subsystems which (i) solve particular, well-defined sub-tasks *by manipulating and/or transforming representations*, and (ii) communicate with each other *by passing representations* (see, e.g., [34]). Criticising my position, Andy Clark [8] rejected (what he called) my "intimate linking of representation-invoking stories to homuncular decomposition", on the following grounds:

> It seems clear that a system *could* be interestingly modular yet not require a representation-invoking explanation (think of a car engine!), and vice versa (think of a connectionist pattern-associator). [8, p.272]

So Clark's objection was that because one can have (A) interesting modularity without representations, and (B) representations without interesting modularity, one should not link homuncular explanation to representational explanation.

These are intriguing complaints, so let's take a closer look at them. It certainly seems beyond doubt that a system *could* be interestingly modular, yet not require a representation-invoking explanation. However, this fact might not have quite the impact that Clark suggests. To claim that a system succumbs to modular *explanation* — where the only modularity of any use will necessarily be of an *interesting* kind — is to claim that the global behaviour of that system can be explained in terms of the collective behaviour of an organised ensemble of identifiable subsystems. This says next-to-nothing about either the specific nature of the subsystems (the modules), or the specific form of any interactions that may occur between them. Thus while *some* modular explanations will not require representation-talk (e.g., the explanation of a driven pendulum in terms of coupled differential equations, of which one describes the motor and one describes the pendulum), this fact leaves plenty of space for a theoretical position according to which (i) homuncular explanations are conceptualised as a subset of modular explanations, and (ii) homuncular explanations always invoke representations.

The reason why Clark thought that his first observation counted directly against my original view becomes clear, once we consider his own definition of homuncular explanation.

> To explain the functioning of a complex whole by detailing the individual roles and overall organization of its parts is to engage in homuncular explanation. [8, p.263]

In effect, this definition collapses the distinction between homuncular explanation and modular explanation. No wonder, then, that from the undeniable fact that one can have interesting modularity without representation, Clark proceeded to infer that one can have homuncularity without representation. However, if I am right, and the conceptual distinction between modular explanation and homuncular explanation is both clear and robust, then any argument which rests on abandoning that distinction must appear suspect.

As it happens, even though Clark's own argument falls short of its intended target, I am now inclined to agree with him that it is a mistake to include internal representation in the definition of homuncular explanation. To see why, we can begin by taking the key features of the homuncular strategy (as identified in section 2), and placing them in the context of the distinction between modular and homuncular explanations. Thus we can state that homuncular explanations are that sub-set of modular explanations in which the modules concerned (i) solve particular, well-defined sub-tasks, (ii) communicate with each other, and (iii) are hierarchically organised. One might, I suppose, be able to construct explanations which meet the conditions for homuncularity, but which do not appeal to the existence of internal representations (in the sense of scientifically well-individuated, content-carrying internal states). Thus the claim that the homuncular strategy presupposes internal representation is (probably) too strong. What remains compelling, however, is the idea that homuncular decomposition *warmly invites* a representational story. The crucial feature here appears to be the presence of inter-module communication. As soon as one thinks in terms of inner modules that *pass messages to each other*, one is, it seems, driven towards thinking of each of those modules as receiving informational inputs from, and sending informational outputs to, other modules. And the natural scientific concomitant to this thought is the idea that the information concerned is internally represented within the system.

We can see how tempting this homuncular-representational package is, if we return, for a moment, to Clark's example of the car engine. A non-homuncular, non-representational, yet uncontroversially modular explanation of a car engine is surely possible. Nevertheless, this fact does not imply (and, presumably, Clark does not intend it to suggest) that one could not have a useful and illuminating homuncular explanation of such a system. It seems obvious that one could. Indeed I transcribed the following tale of woe (almost verbatim) from a real-life conversation.

> My car keeps stalling, which is really annoying. The thermostat is broken, It's telling the automatic choke that the engine is hot enough when, in reality, it isn't. The choke thinks that the engine is hot enough and so it doesn't do anything. That's why the engine stalls.[4]

Bubbling just below the surface of what was actually said here is an homuncular metaphor, in which the thermostat communicates information about the engine-temperature to the automatic choke, which then decides what to do. And once that way of talking has been established, it seems seductively natural to think of the communicated information as being internally represented somewhere within the engine.

Now what about the second prong of Clark's argument: that because one can have internal representations without interesting modularity, one can have internal representations without homuncularity? In contrast to Clark, I have sought to maintain a distinction between modularity and homuncularity. However, even though, on my view, interesting modularity is not sufficient for homuncularity, it is still the case that it is *necessary*, so the conclusion that one can have internal representations without homuncularity would still follow from the fact that one can have internal representations without interesting modularity. Of course, if

---

[4] Many thanks to Sharon Groves who (without even the merest hint of a prompt) produced this little gem of motorised homuncular explanation.

it could be established that no explanatory appeal to internal representations is possible unless one can perform homuncular decomposition on the control system in question, then it would have been shown that one cannot have internal representations without interesting modularity (since any homuncular system must be an interestingly modular system). Under these circumstances, interesting modularity would be necessary but not sufficient for internal representation. But now *can* we establish that homuncular decomposition is necessary for internal representation? Perhaps we can, by appealing to an analysis due to Harvey [15].

Taking natural language to be a paradigmatic case of a representational system, Harvey begins by suggesting that words are essentially arbitrary symbols which receive their meaning from tacit background agreements about how those symbols should be used, agreements which exist between the members of a linguistic community. Generalising this story to representations in general, Harvey proposes the following schema: "A symbol P is used by a person Q to represent, or refer to, an object R to a person S. Nothing can be referred to without somebody to do the referring" [15, p.4]. He later remarks, "when talking of representation it is imperative to make clear who the users of the representation are ... always make explicit the Q and S when P is used by Q to represent R to S" (p.4). Taken as a model for *internal* representation, this account requires the existence of homuncular subsystems within the brain which can act as the metaphorical interpreters for inner representational tokens. In other words, on such a view, it is communities of inner homunculi who use a particular internal state of the agent to *stand in for* a particular feature of the environment during their behaviour-organising communications. To adapt a famous example, if, according to our best theory of the frog's fly-catching-behaviour, homunculus Q is the frog's visual system, and homunculus S is the specific behavioural mechanism responsible for the frog sticking out its tongue, then the content R of one of the frog's inner states P, is "there's a fly over there now", if that's what Q tells S using P.[5]

What all this suggests, I think, is that if one could, with a clear philosophical conscience, adopt a Harvey-esque account of internal representation, then one would have good reason to treat homuncular decomposition as necessary for internal representation. At this point, however, care is needed. As we learned earlier, Clark offers an example — the simple connectionist pattern-associator — which is designed to block the inference from the presence of internal representa-

---

[5] There is an important issue here about how, on this sort of account, to ground the content of internal representations. According to Harvey, the contents of external representations are determined by the background agreements of the community of representation-users, whilst the contents of internal representations (if such things existed) would be intrinsically and unavoidably observer-relative. However, given the naturalism inherent in cognitive science, it seems to me that the more appropriate strategy in the internal case would be to adopt a variation on the popular theme of evolutionary content (e.g., [25]), a variation which made explicit the essential part played by homunculi in the appeal to representations. Further discussion of this issue would, I regret, take us too far afield.

tions to the presence of homunculi; so let's now consider that specific case. If we have in our minds the image of an isolated network, all alone in a void, happily mapping input patterns onto output patterns, then our first reaction is indeed to say that here we have a system which does not succumb to homuncular decomposition, but which does admit of a representational description. In fact, we are being misled. To see why, notice that the connectionist modeller will not be thinking of her pattern-association network as a complete cognitive architecture. Standard connectionist networks are typically trained to execute some functionally well-defined sub-task which has been abstracted away from cognition as a whole (e.g., learning the past-tense of verbs, or transforming lexical input into the appropriate phonological output). This puts the representations realised in Clark's pattern-associator network into context: the sub-task performed by the network will almost inevitably have been conceptualised as a matter of executing transformations between input representations (which, in theory, arrive from elsewhere in the cognitive architecture) and output representations (which, in theory, are sent on to other functionally identified modules). It transpires, then, that although it may *look as if* the representations concerned are not part of an homuncular story, in actual fact they are. It is just that the homunculus who would have been responsible for producing the pattern-associator's input representations, and the homunculus who would have been the beneficiary of the pattern-associator's output representations, plus all the other homunculi in the cognitive architecture, have been assumed/suppressed so that the cognitive scientist can concentrate her attention on the particular sub-task in which she is interested. So it seems that Clark's pattern-association network does *not* demonstrate that one could have internal representations without homunculi; and this, in turn, means that the conceptual space is still available for a position according to which any explanation in cognitive science that includes an appeal to internal representations must also include an (explicit or implicit) appeal to homunculi.[6]

Despite this result, the Harvey-esque view of internal representation is not trouble-free. Here is another problem, and this one seems to ride on the back of the very factor which has seemed to be doing most of the work in connecting homuncular decomposition with internal representations, namely the idea that internal representations are essentially the means by which inner modules *communicate* with each other. The problem here stems from the fact that the very idea of 'representation as communication' implies that it is of the essence of representations that they be *interpreted* or *understood*, by someone or something, before they can have any causal impact on behaviour (for further discussion of this point, see, e.g., [36]). But whilst the requirement of prior interpretation might well be unproblematic in the case of the external, public representations that figure in the communicative dealings of whole intentional agents, there is general agreement that the causal properties of *internal* representations must be

---

[6] Harvey [15] identifies a second way in which connectionist networks may succumb to homuncular decomposition. Given a multi-layered network, one might think of the individual layers within the network as homunculi who communicate with each other.

secured *without those representations first being interpreted or understood*. Parts of the brain engage with internal representations, and, literally speaking, parts of the brain don't understand anything! Of course, as we know, homuncular explanation is served with a large helping of metaphor. But the worry here won't be deflected by the observation that the interpretation of internal representations by inner 'agents' is always metaphorical. No metaphor can be explanatorily useful in science if it misleads us as to the fundamental workings of the target system. After all, the whole point of using metaphors in science is to help make the systems that we investigate intelligible to us. Here, one might think that the homuncular account can still be saved, by its commitment to the principle that the apparently intentional properties of the homunculi should ultimately be discharged by the straightforward non-intentional causation that operates at the lowest level of analysis (see section 2). But I, for one, am no longer convinced that this move works, because there remains, it seems, a mis-match between the brute physical causation apparently doing the discharging, and the interpretation-dependent causation which, as things stand, the homuncular metaphor seems to suggest operates at the higher levels of analysis. Unless the transition between the two sorts of causation can be satisfactorily explained (or explained away), the homuncular metaphor remains, it seems, problematic. And if that is right, and yet it is also true that internal representation depends on homuncularity (as presently understood), then that result would seem to show that the very notion of internal representation is itself suspect.

This entire issue would, of course, fall by the wayside, if we had a satisfactory concept of internal representation which did not buy into the idea of representation as communication. The unfortunate thing about going down this route is that one really would like to give in to temptation, and embrace the compelling thought that where there is an appeal to internal representation, there will be an appeal to intra-system communication between inner modules. Indeed, as soon as one thinks in terms of internal representations, the idea of subsystems that communicate the information that those representations carry becomes difficult to resist, since it seems natural to think of the informational content of those inner states as being essentially *for* inner communications. What, indeed, would be the point of the frog's visual system building a representation indicating the presence of flies, if it didn't then use that state to *inform* the tongue-controlling motor-systems of what was going on?

The beginnings of a resolution to this tension might be found in the following suggestions for how to make our concept of a representational control system more sophisticated.[7] The first step here is to call attention to the fact that for an inner state (one that makes some distinctive causal contribution

---

[7] These suggestions are justified properly, and cashed out in much greater detail, by Andy Clark and I in [36], where they are parts of a much larger machinery with which we attempt to understand a particular class of adaptive solutions to the problem of generating environmentally-embedded behaviour, viz solutions in which intelligent action is produced through the combined causal contributions of multiple factors that are spread across brain, body, and environment. Adaptive solutions of this variety — solutions which involve (what we call) *causal spread* — have seemed to

to adaptive success) to count as an internal representation, it is plausible that that state would have to exhibit a kind of *arbitrariness*, in the sense that the set of different components that could play the same role as that state in the overall behaviour-producing process must be fixed not by the non-informational physical properties of those components, but precisely by their ability to carry specifiable bodies of information. Then, for the information so-carried actually to guide behaviour, what is required, it seems, is the presence of *consumer subsystems* that, although they do not actually understand the information-content of the inner states concerned, can nevertheless be said — *literally* said that is — to exploit those inner states for that information (cf. [25]). This process of content-based consumption might be understood as a kind of inner decoding of the represented information.

How do these ideas help us in the present context? At a first pass, one might be tempted to think that the proposed picture of consuming-decoding inner modules avoids appeals to inner communication altogether. However, if the fully worked out version of the position requires that the information carried by the representational states be produced (or encoded) 'upstream' by other inner modules (again cf. [25]), then a communication-based understanding of the inter-module transactions might be compelling. An alternative way of proceeding, then, might be to develop the idea that although the consumption-decoding picture does involve a notion of communication, it is a notion of communication which does not require the internal representations concerned to present themselves to understanding subsystems as having a certain content. This idea is, I freely admit, in need of further work, but if it has a future (and I think that it has), then the notions of communication and homuncularity that we need, in order to preserve the inference from internal representation to homuncularity, might well be substantially leaner than we had previously supposed.[8]

The final worth of the conceptual modifications suggested here will, of course, be decided only by the results of further philosophical and empirical research. Nevertheless, we must surely have enough evidence to conclude that the idea of internal representation is *at least* strongly suggestive of the notion of homuncular decomposition. Thus, given our earlier observations, I think we can justifiably treat it as established that there exists, in cognitive science, a relation of mutual support between, on the one hand, explanations that appeal to internal representations and, on the other, the homuncular decomposition of the systems that we wish to explain.

We now have just about as firm a grip as we are going to get (for this paper at least) on these issues. It is time, then, to bring evolutionary robotics into the game.

---

many theorists to be resistant to explanations in terms of internal representation. However, we defend a notion of internal representation that (we think) does retain explanatory power in such cases.

[8] I should mention here that in [36], Clark and I argue that the consumption-decoding view does not buy into the idea of representation as communication at all. However, in that work, we consider only the strong, interpretation-requiring, undoubtedly problematic version of that idea.

4 Will Evolution Play Ball?

To have empirical clout (i.e., to be anything more than a nice idea), the homuncular strategy requires that biological agents be seen to embody a certain sort of scientifically identifiable neural modularity, one that is neat, hierarchical, and involves internal communications. In addition, given the conceptual ties that, as we have seen, exist between homuncular decomposition and internal representation, homuncular explanation predicts (i) that there exist structures in the agent's brain that the cognitive scientist can usefully interpret as internal representations, and (ii) that the communications between neural modules will usefully be understood as processes in which such structures are employed. But now who's to say that evolved systems will carve-up the way we want them to?

To appreciate the very real challenge laid down by this question, we need to shift domains for a moment, and visit the world of engineering. In general, engineers work with well-specified problems, and engineering solutions reflect the designer's functional conceptualisation of the task at hand. Thus when an engineer begins the job of designing a system to solve a complex problem, her standard tactic will be to analyse that complex problem into a number of simpler, well-defined, inter-related sub-problems. Typically, the structure of the final system will respect that task-decomposition, so the end result will be a collection of functionally identified, communicating subsystems, organised into some sort of overall control-system hierarchy. This methodology — which is, of course, tantamount to homuncular decomposition — is most visible in the design of commercial software systems, in which functionally specified, hierarchically organised modules are built to carry out well-defined computations, and to communicate with each other via representations.

The application of homuncular-style thinking within engineering has proved itself to be extraordinarily powerful, enabling the human designer to overcome a vast range of problems. However, as many evolutionary roboticists are inclined to observe, that same strategy might well be found wanting, when applied to the problem of building control systems which are capable of generating environmentally embedded adaptive behaviour. Broadly speaking, there are two reasons for this scepticism, and these reasons reconnect us with the biological realm:[9]

1. From an evolutionary perspective, the fundamental goal of an animal is to survive long enough to reproduce. In a noisy, dynamic, and possibly hostile environment, the full suite of ecological constraints on achieving this goal will be difficult (perhaps even impossible) to specify. Moreover, the widespread presence of coevolutionary situations means that the adaptive problems which the creature faces will typically be subject to evolutionary change. If the artificial autonomous agents that we build are to be embedded

---

[9] Here I am merely reconstructing arguments of a kind relatively common in evolutionary robotics circles.

in environments which are even remotely similar to the environments of animals — in terms of dynamic uncertainty — then the problems confronted by those artificial agents will also be difficult to specify and unavoidably open-ended. Thus any attempt to identify a set of well-defined tasks and sub-tasks for putative inner homunculi to perform may be doomed to failure.

2. Natural selection acts merely to retain the control systems of those creatures which consistently survive long enough to reproduce. The basic constraint on the design of a biological control system is thus that it allows the animal in question to generate the adaptive behaviour which will enable it to achieve that goal. There is no additional constraint to the effect that the organisation of that control system must embody the style of decomposition traditionally favoured by human designers. Thus it cannot be taken for granted that biological nervous systems will realise homuncular decomposition.

Enter, of course, *evolutionary robotics*, in which the role of the human designer (with all her predilections for homuncular-style control systems) is usually reduced to a practical minimum.

Just how much input into the evolutionary process the human designer should have is itself an important question. One design decision that, as things stand, does remain her responsibility is the choice of building blocks to be used by artificial evolution. In the evolution of adaptive control systems, there are a number of reasons why artificial neural networks emerge as natural and promising candidates for this role. As Husbands et al. [21] explain, these include (a) the smoothness of the fitness landscapes that neural networks generate (a phenomenon which aids evolvability), (b) the suitability of neural networks for ongoing open-ended evolution (in which evolution will continue to produce better adaptive solutions, if better adaptive solutions exist, and so must be able to increase or decrease system-complexity — see [22]), and (c) the fact that neural networks provide evolutionary primitives of an appropriately low level (so that, to a large extent, the introduction of human preconceptions about how the final system ought to work can be avoided). But now what *sort* of network should be used? The types of network familiar from most mainstream connectionist research are, according to some theorists, unsuitable, due to the fact that those networks feature certain deliberately-introduced architectural restrictions, such as (i) neat symmetrical connectivity, (ii) activation passes that proceed in an orderly feed-forward fashion from an input layer to an output layer, (iii) noise-free processing, (iv) units which are uniform in structure and function, and (v) timing properties which are based either on a global, digital clock (which keeps the progressive activity of the units in synchronization) or on methods of stochastic update [16]. Such restrictions, it is argued, limit the range of possible network dynamics available for evolution to exploit. Thus, in the interests of exploring the possibilities presented by richer control system dynamics, some evolutionary roboticists favour so-called *dynamical neural networks* (henceforth *DNNs*). In addition to being highly distributed systems (a quality that they share with many of their more conventional relations) DNNs feature the following sorts of properties: asynchronous continuous-time processing, real-valued time delays on connections, non-uniform activation functions, deliberately introduced noise, and connectivity which is not only both directionally unrestricted and highly recurrent, but also not subject to symmetry constraints. Networks featuring various sub-sets of these properties are deployed by, for example, Beer and colleagues (see, e.g., [2, 3]) and by members of the Sussex evolutionary robotics group (see, e.g., [10, 17, 20, 21]).[10]

So what happens when artificial evolution is given the task of designing DNNs to generate adaptive behaviour in artificial autonomous agents? The existing evidence from the ongoing work at Sussex will be well-known to many readers of this paper, but it bears repeating. In the Sussex studies using DNNs, the general commitment to the principle that as few restrictions as possible should be placed on the potential structure of the network manifests itself as a 'hands-off' regime in which artificial evolution is allowed to decide such fundamental architectural features as the number, directionality, and recurrency of the connections, and the number of internal units. In addition, certain aspects of the robot's visual morphology are (typically) placed under evolutionary control. Under this regime, robots with DNN controllers have been successfully evolved to carry out relatively simple homing, discrimination, and target-tracking tasks.[11] For present purposes, the crucial fact about the evolved controllers is that even after the DNNs concerned have been subjected to a preliminary simplifying analysis — in which (a) redundant units and connections (which may have been left over from earlier evolutionary stages) are eliminated, and (b) the significant visuo-motor pathways are identified — it is still the case that the salient channels of activation connecting sensing with action flow in highly complicated and counter-intuitive ways, due to the complex connectivity and feedback that are present within the networks (for the details, see, e.g., [10, 17, 20]). This dynamical and structural complexity has suggested, to some theorists, that such systems will, in general, be resistant to representational analysis. At a first pass, this is because, as Beer comments, "highly distributed and richly interconnected systems [such as

---

[10] Once DNNs are employed as robot controllers, the use of artificial evolution as a design strategy might well be (near enough) mandatory. The thought here is that if DNNs are to be exploited to their full potential, then the systems that we wish to (somehow) organise appropriately will be highly unconstrained, in that they will feature large numbers of free parameters (e.g., time delays, activation functions). Such systems would be extraordinarily difficult — perhaps even impossible — for humans to design (see [21, pp.135-6]). And if this is right, then, given of course that artificial evolution does indeed produce successful controllers from such systems, then the intuition (widely held amongst evolutionary roboticists), that artificial evolution can explore a wider space of possible control architectures than could any human being, would be vindicated.

[11] It is interesting that many of these controllers exploit the phenomenon of causal spread (see footnote 7), in order to achieve adaptive success. Evolutionary processes are, it seems, encouraged to find this style of adaptive solution in cases where the sensor morphology is evolved concurrently with the DNN. A specific example of an evolved DNN-based control system that features causal spread (a control system evolved during the experiments described in [17]), is discussed in [36].

evolved DNNs] ... do not admit of any straightforward functional decomposition into representations and modules which algorithmically manipulate them" [1, p.128]. Beer's remark undoubtedly points us in the right direction. However, it would be useful if we could go further, and get a proper grip on exactly *why* and *how* representational explanation is undermined by such systems. To do this, we can plug-in some general considerations about certain styles of causal system.

Let's begin by observing (along with Clark [9, p.114]) that the generic class of causal systems for which homuncular explanation works best is the class of what Wimsatt [37] calls *aggregate systems*. (For reasons that should become clear soon, this is true on either Clark's or my understanding of homuncularity.) An aggregate system is a system in which (a) it is possible to identify the explanatory role played by any particular part of that system, without taking account of any of the other parts, and (b) interesting system-level behaviour is explicable in terms of the properties of a small number of parts. A system becomes progressively less aggregative as the extent and complexity of the interactions between its parts increases. Clark identifies the causation that leads to such non-aggregativity as *continuous reciprocal causation* [7, 9], defined as causation that involves multiple simultaneous interactions and complex dynamic feedback loops, such that (a) the causal contribution of each component in the system partially determines, and is partially determined by, the causal contributions of large numbers of other components in the system, and, moreover, (b) those contributions may change radically over time.[12] Using this terminology, we can state that the aggregativity of a system is negatively correlated with the degree of continuous reciprocal causation within that system. Now for the key move. It seems plausible that, as a system becomes less and less aggregative, with increasing continuous reciprocal causation, the most useful explanatory stance that one can take towards that system will become increasingly holistic, i.e., the most useful explanations will become increasingly *non-modular*. The justification for this claim is that the sheer number and complexity of the causal interactions in operation in such systems force the grain at which useful explanations are to be found to become coarser. So our explanatory interest will be compulsorily shifted away from the parts of the system and their interrelations — and therefore, eventually, away from any modular analysis — and towards certain 'higher-level' system dynamics.[13]

---

[12] It is important to realise that continuous reciprocal causation and causal spread (see footnote 7) are conceptually distinct phenomena, even if, as a matter of empirical fact, they are often found together. This conceptual separability is indicated by the fact that one could have a system in which causal spread is rife, but in which the multiple causal factors concerned play distinct roles and combine linearly to yield action.

[13] I take it that the modular explanation of a system can be undermined by continuous reciprocal causation, *even where that phenomenon does not affect the entire system*. Imagine, for example, a complex system in which the expansion of continuous reciprocal causation is restrained, in such a way, however, that large regions of that system provide the only candidates for explanatorily useful 'inner modules'. It seems that any attempt to construct a useful modular explanation of this system would al-

It is a crucial part of this picture that even large-scale continuous reciprocal causation does not necessarily put the system in question out of our explanatory reach. One might, for example, pursue a particular style of dynamical systems analysis, in which the systemic variables that matter to the explanation do not correspond straightforwardly to the properties of internal components, but rather to features of a higher-level state space (see [30]). Something like this tactic is adopted by Husbands et al., in an analysis where a low-dimensional state space is used — in conjunction with a method by which evolved DNN dynamics are collapsed into the properties of certain feedback loops — to understand how the evolved behaviours of a simple robot vary with the visual input which that robot receives [20].[14]

We can now combine our appreciation of aggregativity with our earlier analysis of the link between representational and homuncular forms of explanation, to state more precisely why the kind of evolved DNN control system under consideration might be resistant to a representational understanding. It seems likely that networks of this sort — characterised, as they are, by complex connectivity and dynamic feedback — will consistently exhibit high degrees of continuous reciprocal causation; so (typically) those networks will be non-aggregate systems. Given the foregoing analysis, this suggests that the most effective explanatory stance which one could take towards such systems will (typically) be one which is holistic rather than modular. *But now homuncular explanations are a subset of modular explanations. So homuncular explanations will (typically) not be useful in such cases.* This leads us to the anti-representational conclusion: to the extent that homuncular explanation and representational explanation are mutually supporting, any threat to the former will apply equally to the latter. Thus, given the threat to homuncular explanation that we have just identified, we have, it seems, good reasons (of a rather general kind) to think that the concept of internal representation will not be useful in helping us to understand evolved DNN control systems.

The keystone of the anti-representational argument is the claim that any modular explanation of evolved DNN controllers will typically be blocked by the high degrees of continuous reciprocal causation observed in those systems. With this in mind, let's now consider studies by Thompson [29], where, still with the goal of producing adaptive control systems for autonomous robots, artificial evolution is applied not to artificial neural networks, but to the even lower-level evolutionary primitives offered by reconfigurable hardware. Thompson highlights two constraints that are standardly placed on electronic circuits, constraints that are imposed with the aim of rendering those circuits amenable to human design. The first is familiar to us: it is the "modularisation of the design into parts

---

ready have been undermined, since, given the coarseness of the identifiable modules, much of the flexibility, adaptiveness, and behavioural richness of the system would, from any modular explanatory perspective, remain a mystery.

[14] Here is not the place to launch into a detailed discussion of the use of dynamical systems explanations in evolutionary robotics (see [1, 2, 16, 20]). For philosophical reflections on this issue, see [34, 35].

with simple, well defined interactions between them" [29, p.645]. The second is the inclusion of a clock: this gives the components of the system time to reach a steady state, before they affect other parts of the system. Thompson argues that, once artificial evolution is brought into play, both of these constraints should be relaxed, since the richer intrinsic control-system dynamics that will result might well be exploited by evolution, even though human designers are unable to harness them (cf. the relaxing of restrictive architectural assumptions in the case of DNNs). However, to have any chance of exploiting the dynamical possibilities presented by abandoning the controlling clock, one has to overcome a problem, namely that electronic components usually operate on time scales which are too short to be of much use to a robot. So one would like artificial evolution to be capable of producing a system in which, without there being any clock to control different time-scales, the overall behaviour of a whole network of components is much slower than the behaviour of the individual components involved. Thompson set artificial evolution this task, using, as his evolutionary raw material, a population of (simulated) recurrent asynchronous networks of high speed logic gates. After forty generations (when the experiment was called to a halt even though fitness was still rising), a network had evolved which produced output that was over four thousand times slower than that produced by the best of the networks from the initial random population, and six orders of magnitude slower than the propagation delays of the individual nodes. What is striking is that the successful network seems to defy modular decomposition. As Thompson reports, the "entire network contributes to the behaviour, and meaningful sub-networks could not be identified" (p.648). Thus we have a powerful example of artificial evolution producing a non-aggregate system, a system where (a) the complex nature of the causal interactions between the components means that "meaningful sub-networks [interesting modularity, functionally discrete subsystems] could not be identified", and (b) the system has to be understood in an holistic manner ("the entire network contributes to the behaviour").

To be convinced that these results have any potentially important consequences for cognitive science, one first has to accept that there is a sufficient degree of similarity between the artificially evolved control systems concerned and their natural counterparts. There is little doubt that the generic architectural properties that characterise DNNs (recurrency, noise, real-time dynamics etc.) reflect, to some extent, the generic architectural properties of biological nervous systems. Despite these affinities, however, we need to be on constant guard against drawing any overly enthusiastic, misleading parallels. The most telling observation, I think, is that the structural complexity observed in many existing examples of evolved DNN controllers resonates with the structural complexity of real biological nervous systems, in that some of the features that artificial evolution discovers in these cases have analogues in real nervous systems (e.g., the repeat-pattern generators discovered in evolutionary robotics [2, 23], and in neurobiology [4]).

Thompson's experiments in evolutionary electronics are suggestive in this context, because the results indicate that where evolutionary processes have access to properties of the hardware, those processes will produce control systems whose intrinsic physical dynamics are crucial to adaptive success (for the details, see [29]). This promotion of certain 'low-level' physical dynamics provides a second bridge to biological systems, because it is an important but far-too-often forgotten fact that, whatever else they may be, biological brains are undoubtedly complex *chemical* machines.[15] For instance, there is experimental evidence that when the neurotransmitter glutamate is presynaptically released in the brain, it propagates in clouds. Because of this, presynaptic releases of glutamate may affect not only 'their own' post-synaptic receptors, but also, through a sort of spill-over phenomenon, other, distant post-synaptic receptors [24]. There is no reason to suppose that chemical effects such as glutamate spill-over — effects which are certainly beyond the standard connectionist picture of signals sent along neatly specifiable pathways — can be ignored as unimportant 'implementation details' or as 'noise' in our cognitive-scientific investigations. Rather, just as the hardware dynamics in Thompson's artificially evolved systems are essential contributors to adaptive success, there is every possibility that chemical dynamics are essential features of the 'strategies' by which naturally evolved brains generate biological adaptive behaviour.

The dual themes of network complexity and physical dynamics are brought together in some of the most recent work in evolutionary robotics, work which seems to take us even further from the standard cognitive-scientific framework, and which might well be interpreted (with some justification I think) as promoting an increasingly biologically-oriented approach within evolutionary robotics. Husbands [19] has successfully evolved artificial control systems (guiding simple robot behaviours) in which there are a number of distinct but interacting dynamical mechanisms, analogous to the electrical, short range chemical and long range diffusing chemical processes in real nervous systems. In Husbands' artificial systems the 'electrical' and the 'chemical' processes are deeply intertwined. Diffusing 'clouds of chemicals' can change the intrinsic properties of the artificial neurons, and trigger emissions of other 'chemicals', whilst 'electrical' activity can trigger 'chemical' activity. This intertwining reflects a certain contemporary view of how biological brains turn their adaptive tricks, a view which stresses the interaction between signal-processing and chemical dynamics. (For discussion of this view, see [5].)

Perhaps the most that one can say with complete confidence is that the general dynamical properties realised by many existing examples of evolved artificial control systems bear interesting resemblances to the general dynamical properties realised by biological nervous systems. However, this still means that one can — with great care and due hesitancy — make well-supported claims about how biological systems might work, on the strength of appropriate experiments in evolutionary robotics. Moreover, the number of opportunities to make such claims might well be on the increase. With this, the argument targeted at the beginning of this paper is finally in place. It seems that there is reason to believe that certain results in evolutionary robotics pose a threat to traditional homuncular-representational forms of explanation in cognitive science.

---

[15] Thanks to Mick O'Shea for bringing this issue into focus for me.

5 The Shape of Things to Come

Having travelled this far with me, many readers will be wanting to ask a rather obvious question: what will happen when evolutionary roboticists manage to evolve autonomous agents with behavioural capacities (or, perhaps even more to the point, suites of behavioural capacities) which are significantly more complex than those that have been evolved so far? Will they end up rediscovering control architectures that are usefully explained as being modular, homuncular, and representational? Resisting the temptation to engage in runaway speculations here, I shall offer a few brief and sketchy remarks concerning two potential misunderstandings that might prevent us from keeping in view all the possible answers to such questions. These remarks focus on issues of modularity, since even though not all modular explanations are homuncular explanations, modularity is necessary for homuncularity, and it is the modular decomposition of evolved control systems which is threatened directly by the presence of continuous reciprocal causation in such systems.

The first of these potential misunderstandings is revealed if we consider some improvements that need to be made to the nuts and bolts of evolutionary design. Just about everyone involved in evolutionary robotics agrees that to evolve control systems which are significantly more complex than those that exist already, genetic encoding schemes will be needed that go beyond the present norm of directly describing the entire network wiring in the genotype. What will be required, among other things, are genetic encoding schemes that, as in nature, allow the building of multiple examples of a particular phenotypic structure, in a single 'organism', through a developmental process in which sections of the genotype are used repeatedly [21]. At first sight, this might seem to wrap up the issue of control-system modularity: the advent of richly modular genotypic encodings will lead to robust modularity at the phenotypic, control-system level, and all that worrying about the lack of modularity in existing evolved DNNs, and in existing systems with evolved hardware, will turn out to have been nothing more than a bad dream. In fact, the issues here are surely much less clear-cut than this reasoning implies. The question that matters for the modular *explanation* of control systems is not 'will a modular genetic encoding scheme produce repeated structures in the evolved control system?', to which the answer is presumably a trivial 'yes'. Rather, it is 'will such repeated structures causally contribute to adaptive success in ways (i) which are not determined (to an overwhelming extent) by the causal contributions of large numbers of other structures in the system, and (ii) which do not change radically over time?', to which the answer is, I think, 'we simply don't know — not yet'.

At this point, the second potential misunderstanding is liable to kick in. "Surely", it will be said, "your open-mindedness is misguided, because current neuroscience has established that the brains of most animals already succumb to modular explanation, showing beyond doubt that such systems do not suffer from the insidious effects of continuous reciprocal causation". As tempting as this conservatism might be, we have to remind ourselves of the (hermeneutical) fact that any particular scientific explanation of a complex system is something which is not entirely neutral with respect to the set of preconceptions which the theorist brings to the phenomena under investigation. Indeed, it is the interaction between, on the one hand, the theorist's preconceptions, and, on the other, the understanding of the system thereby obtained, that makes one proposed explanation more compelling than another. At the very least, then, it is possible that we might actually be *restricting* our understanding of the way complex biological nervous systems *often* work, by assuming that *all* our explanations of neural phenomena must be modular in form. Given this, one might (as I do) find it suggestive that Varela, Thompson and Rosch [32] have developed and defended an account of neural processing which seems both to recognise the phenomenon of continuous reciprocal causation, and, on the strength of that recognition, to urge us towards a more holistic form of explanation.

> [One] needs to study neurons as members of large ensembles that are constantly disappearing and arising through their cooperative interactions and in which every neuron has multiple and changing responses in a context-dependent manner ... The brain is thus a highly cooperative system: the dense interconnections among its components entail that everything going on will be a function of what all the components are doing ... [If] one artificially mobilizes the reticular system, an organism will change behaviorally from, say, being awake to being asleep. This change does not indicate, however, that the reticular system is the controller of wakefulness. That system is, rather, a form of architecture in the brain that permits certain internal coherences to arise. But when these coherences arise, they are not simply due to any particular system. (p.94)

Here is not the place to embark on a detailed discussion of the specific framework defended by these authors. I mention the work only to make the general point that scientifically respectable, serious alternatives to the standard ways of thinking about neural systems do exist. Evolutionary robotics may well be crucial in helping us to explore those alternatives.

6 Conclusions

My goals have been modest ones. I have tried to improve our intellectual grip on the nature of representational explanatory strategies in cognitive science; and I have tried to articulate, in general terms, a challenge to such strategies, one which is apparently mandated by certain results in evolutionary robotics. In a way, this paper has been about why it is hard to explain a certain category of evolved control systems using the traditional explanatory language of cognitive science. Whether or not increasingly sophisticated evolved control systems will be equally as recalcitrant is, I have argued, much more of an open question than many theorists might imagine.

Acknowledgements

This work was supported by a Junior Research Fellowship at Christ Church, Oxford, with additional assistance from the McDonnell-Pew Centre for Cognitive Neuroscience, Oxford. Many thanks to Maggie Boden, Andy Clark, Martin Davies, Inman Harvey, Susan Hurley, Ronald Lemmen, Tim Smithers, and Tim van Gelder for invaluable discussions which have taken place at various times during the development of the ideas presented here, and to Phil Husbands for equally invaluable discussions and for comments on an earlier version of the paper.

References

1. R. D. Beer. Computational and dynamical languages for autonomous agents. In [27], 121-47, 1995.
2. R. D. Beer. A dynamical systems perspective on agent-environment interaction. *Artificial Intelligence*, 72:173–215, 1995.
3. R. D. Beer and J. G. Gallagher. Evolving dynamic neural networks for adaptive behavior. *Adaptive Behavior*, 1:91–122, 1992.
4. P.R. Benjamin and C.J. Elliot. Snail feeding oscillator: the central pattern generator and its control by modulatory interneurons. In J.W. Jacklet, editor, *Neuronal and Cellular Oscillators*, chapter 7, pages 173–214. Dekker, 1989.
5. M H. Bickhard and L. Terveen. *Foundational Issues in Artificial Intelligence and Cognitive Science: Impasse and Solution.* Elsevier, Amsterdam, 1996.
6. M. A. Boden, editor. *The Philosophy of Artificial Life.* Oxford University Press, Oxford, 1996.
7. A. Clark. Twisted tales: Causal complexity, dynamics and cognitive scientific explanation. Forthcoming in *Minds and Machines.*
8. A. Clark. Happy couplings: Emergence and explanatory interlock. In [6], 262-81, 1996.
9. A. Clark. *Being There: Putting Brain, Body, and World Together Again.* MIT Press / Bradford Books, Cambridge, Mass. and London, England, 1997.
10. D. Cliff, I. Harvey, and P. Husbands. Explorations in evolutionary robotics. *Adaptive Behavior*, 2:73–110, 1993.
11. D. Cliff, P. Husbands, J.-A. Meyer, and S. W. Wilson, editors. *From Animals to Animats 3: Proceedings of the Third International Conference on Simulation of Adaptive Behavior*, Cambridge, Massachusetts, 1994. MIT Press / Bradford Books.
12. D. Cliff and J. Noble. Knowledge-based vision and simple visual machines. *Philosophical Transactions of the Royal Society: Biological sciences*, 352 (1358):1165–75, 1997.
13. D. Dennett. *Brainstorms.* Harvester Press, Sussex, 1978.
14. J. A. Fodor. *Psychosemantics: The Problem of Meaning in the Philosophy of Mind.* MIT Press / Bradford Books, Cambridge, Massachusetts, 1988.
15. I. Harvey. Untimed and misrepresented: Connectionism and the computer metaphor. Cognitive Science Research Paper 245, University of Sussex, 1992. Reprinted in *AISB Quarterly*, 96, pp.20-7, 1996.
16. I. Harvey. *The Artificial Evolution of Adaptive Behaviour.* PhD thesis, School of Cognitive and Computing Sciences, University of Sussex, 1994.

17. I. Harvey, P. Husbands, and D. Cliff. Seeing the light: Artificial evolution, real vision. In [11], 392-401, 1994.
18. H. Hendriks-Jansen. *Catching Ourselves in the Act: Situated Activity, Interactive Emergence, Evolution, and Human Thought.* MIT Press / Bradford Books, Cambridge, Mass. and London, England, 1996.
19. P. Husbands. Evolving robot behaviours with diffusing gas networks. This volume.
20. P. Husbands, I. Harvey, and D. Cliff. Circle in the round: State space attractors for evolved sighted robots. *Robotics and Autonomous Systems*, 15:83–106, 1995.
21. P. Husbands, I. Harvey, D. Cliff, and G. Miller. Artificial evolution: A new path for artificial intelligence? *Brain and Cognition*, 34:130–59, 1997.
22. N. Jakobi. Encoding scheme issues for open-ended artificial evolution. In H-M. Voigt, W. Ebeling, I. Rechenberg, and H-P. Schwefel, editors, Proceedings of *Parallel Processing in Nature*, pages 52–61, Berlin and Heidelberg, 1996. Springer-Verlag.
23. N. Jakobi. Minimal simulations for evolutionary robotics. Submitted DPhil Thesis. School of Cognitive and Computing Sciences, University of Sussex, 1998.
24. D.M. Kullman, S.A. Siegelbaum, and F. Aszetly. LTP of AMPA and NMDA receptor-mediated signals: Evidence for presynaptic expression and extrasynaptic glutamate spill-over. *Neuron*, 17:461–74, 1996.
25. R.G. Millikan. *White Queen Psychology and Other Essays for Alice.* MIT Press / Bradford Books, Cambridge, Mass. and London, England, 1995.
26. F. Moran, A. Moreno, J.J. Merelo, and P. Chacon, editors. *Advances in Artificial Life: Proceedings of the Third European Conference on Artificial Life*, Berlin and Heidelberg, 1995. Springer-Verlag.
27. R. Port and T. van Gelder, editors. *Mind as Motion: Explorations in the Dynamics of Cognition.* MIT Press / Bradford Books, Cambridge, Mass., 1995.
28. E. Thelen and L. B. Smith. *A Dynamic Systems Approach to the Development of Cognition and Action.* MIT Press, Cambridge, Mass., 1993.
29. A. Thompson. Evolving electronic robot controllers that exploit hardware resources. In [26], 641-57, 1995.
30. T. van Gelder. Connectionism and dynamical explanation. In *Proceedings of the Thirteenth Annual Conference of the Cognitive Science Society*, pages 499–503, 1991.
31. T. van Gelder. What might cognition be if not computation? *Journal of Philosophy*, XCII(7):345–81, 1995.
32. F. J. Varela, E. Thompson, and E. Rosch. *The Embodied Mind: Cognitive Science and Human Experience.* MIT Press, Cambridge, Mass. and London, England, 1991.
33. B. Webb. Robotic experiments in cricket phonotaxis. In [11], 45-54, 1994.
34. M. Wheeler. From activation to activity: Representation, computation, and the dynamics of neural network control systems. *AISB Quarterly*, 87:36–42, 1994.
35. M. Wheeler. From robots to Rothko: the bringing forth of worlds. In [6], 209-36, 1996.
36. M. Wheeler and A. Clark. Genic representation: Reconciling content and causal complexity. Submitted for Journal Publication.
37. W. Wimsatt. Forms of aggregativity. In A. Donagan, N. Perovich, and M. Wedin, editors, *Human Nature and Natural Knowledge*, pages 259–93. Reidel, Dordrecht, 1986.

Some Problems (and a Few Solutions) for Open-Ended Evolutionary Robotics

Nick Jakobi[1] and Matthew Quinn[1,2]

School of Cognitive and Computing Sciences[1],
Centre for Computational Neuroscience and Robotics[2],
University of Sussex, Brighton, BN1 9SB, UK,
nickja@cogs.susx.ac.uk,
matthewq@cogs.susx.ac.uk

Abstract. Many of the techniques commonly used to evolve neural networks for robots face various problems if the size and morphology of these networks are allowed to vary under evolutionary control. This paper identifies some of these problems for commonly used encoding schemes, neural networks, genetic operators and evaluation techniques and proposes some new techniques that may be used to alleviate them. Results from experiments are reported in which these new techniques were combined for the open-ended evolution of a many-faceted robot behaviour.

1 Introduction

Much of the Evolutionary Robotics work that has been undertaken to date concerns the evolution of fixed-architecture neural networks for the control of robots [3, 19, 17]. However, it is also possible to evolve neural networks for robot control whose size and morphology are under the control of the evolutionary process itself [10, 7]. While neither approach is definitive, we believe that when evolving large neural networks, capable of displaying complex behaviours, placing the size and morphology of the architecture under evolutionary control has several advantages:

- Fixing the size and morphology of the architecture beforehand places a limit on the levels of complexity that may, potentially, evolve.
- When using a fixed architecture to evolve a complex behaviour, the size and morphology of a sufficient network can only be estimated by the designer. If this fixed architecture is too small or the morphology is inadequate, then the behaviour cannot evolve. If it is too large, then the size of the search space is increased unnecessarily. Putting the size and morphology of the network under evolutionary control overcomes this problem.
- The evolution of a complex behaviour may often involve the cumulative evolution of many behavioural sub-modules in an incremental fashion. If this is undertaken using a large, fixed architecture, then evolution must deal with the whole network at each step. If the size and complexity of the network is under evolutionary control, then only the necessary mechanism for the

sub-module and its integration into the rest of the network needs to evolve at each step.

This paper examines some of the problems, specific to Evolutionary Robotics, that occur if the size and morphology of controllers are allowed to evolve in an open-ended fashion. It also proposes some solutions to these problems. In Section 2 we discuss how the way in which genotypes encode controllers must be robust to variation in genotype size, and how the simple and commonly used direct encoding scheme fails in this regard. Alternative schemes have been proposed which *are* robust to variable size genotypes but these are much more complicated and entail greater computational and programming overheads. The remainder of the section puts forward a simple encoding scheme that is only marginally more complex than direct encoding but which is robust to variable size genotypes. Section 3 looks at some of the problems for the functionality of the neural networks themselves if their connectivity is allowed to vary. A new type of network is introduced which, while it has its limitations, overcomes many of these problems. In Section 4, the implications of variable size genotype are examined for the evolutionary machinery itself, and the commonly used crossover and mutation operators are shown to be insufficient. Suggestions for how this might be remedied are offered in the form of a new type of mutation operator. The problems specific to the evaluation process are examined next in Section 5. Since open-ended evolution may go on for an arbitrary length of time, the need is identified for a method that is, above all, fast, and a brief sketch is drawn of how one might go about designing and building fast-running minimal simulations on which these evaluations may be performed. In section 6, details of experiments are reported in which all of the new techniques and methods proposed in the paper are combined in the open-ended evolution of a complex robot behaviour. Finally Section 7 offers some conclusions.

2 Encoding-scheme problems

If we want to do open-ended evolution where the complexity of a solution is not known beforehand then we need to use an encoding scheme that allows the neural networks that it encodes to vary in size. Such a scheme must enable neurons to be added or subtracted from a network without overly disrupting its shape and form. Furthermore, if we are using crossover, then this must be able to produce viable offspring when performed on two genotypes that code for networks of different size. This is actually quite hard to achieve since variation in genotype length means variation in the relative positions of each neuron's coding section on the genotype. Thus a simple direct encoding scheme which specifies the links between neurons by their coding positions on the genotype (e.g. neuron 4 connects to neuron 2) will be substantially disrupted by changes in genotype length. Cliff, Harvey and Husbands suggest as a solution using a mixture of absolute and relative addressing [1], but this is still subject to similar problems. Others have tried to avoid the issue altogether by introducing mechanisms that allow the size of the network to vary while keeping the size of the genotype fixed [12, 2]. However, these schemes come with their own limitations not least of which is that they set an upper limit on the amount a neural network can grow which may not always be desirable.

Gruau has put forward an encoding scheme for the evolution of neural networks called 'cellular encoding' [5] which has been applied successfully to the evolution of controllers for a simulated hexapod robot [6] and a real octopod robot [7]. This scheme is extremely robust to variation in genotype size but it is complicated and unintuitive, and in order to implement it properly entails considerable computational and programming overheads. There are certain things to which it is uniquely suited, such as generating repeated structure, but if the problem at hand does not require this, then the simpler and more intuitive spatially determined encoding described below may be more suitable.

2.1 Spatially determined encoding

In a conventional direct encoding scheme, the genotypic code for a neuron and its links has a particular location on the genotype. The code for each connection specifies another location on the genotype (either through relative or absolute addressing) at which the code for the target neuron of that particular connection can be found. As stated above, this is unproblematic for fixed-size genotype. However, when genotypes are allowed to vary in size, both the relative and absolute positions of the genes encoding each neuron can change - this can have massive and highly disruptive effects on both the shape and functionality of the network.

With a spatially determined encoding, each neuron also has a location, and links between neurons are also specified using these locations. However, in a spatially determined encoding these are locations within a virtual developmental space that is uncorrelated with the spatial organisation of the genotype. The code for each neuron not only specifies its thresholds, time constants and so on, but also its location in this developmental space, and links are specified using these locations. The advantage of this is that it does not matter where the code for a neuron is situated on the genotype, its location within the virtual space remains the same, and thus so do its links to and from other neurons. The idea of spatially locating each neuron has also been employed by other more complex developmental encoding schemes[11, 18]. The simple spatially determined encoding scheme in the form described below was first used by Jakobi in [14]

The left-hand side of figure 1 shows a two-dimensional example of a typical virtual space in which development can take place. For each possible network input and output, a region of this space is specified. Development begins by plotting the position of each neuron into the space according to its genetically specified location. Depending on which region each neuron happens to fall within it is identified as an input neuron, an output neuron or a hidden neuron.

After plotting the locations of all of the neurons in the network, the connectivity of the network is then determined. A large part of the genetic code for each neuron specifies a certain number of links both to and from other neurons.

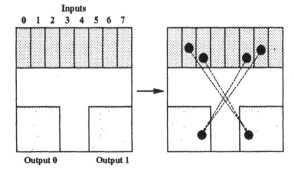

Fig. 1. A simple network generated using the spatial encoding scheme. The left-hand diagram shown how the developmental space can be divide up into input regions, output regions, and 'hidden' unit regions. The eight inputs in this example might correspond to the IR sensors of a khepera robot and the outputs to its two motors. The network on the right, typical of those that might underly obstacle avoidance in a khepera, is generated through reflection from the genotypic specification of three neurons.

The code for each link specifies a target location within the virtual space. If there are any neurons located within a predefined range of this target location, then the nearest one of these is identified as the neuron which the link connects to. If there happen to be no neurons situated within this range then the link fails to connect. The right-hand side of figure 1 shows a very simple, symmetrical network developed from the genetic code for three neurons by a process of reflection.

As seen, spatially determined encoding is simple and yet it suffers none of the problems that afflict direct encoding schemes when genotypes are allowed to vary in size. Crossover between genotypes of different lengths (if it is required, see Section 4) may also be easily achieved through reference to the developmental space. Instead of randomly determining a crossover point on the genotype, a crossover line is determined on the virtual space. If two parent genotypes of different length are selected to form a single offspring genotype, then the sections of parent 1 that code for neurons which are located in the virtual space on one side of the crossover line are joined to those sections of parent 2 that code for neurons which are located on the other side to form the offspring[1]. In this way the crossover operation, like the encoding scheme, is independent of the genotypes' spatial organization[2], and therefore does not run into any of the problems associated with crossover between genotypes of different lengths.

---

[1] Note that this can easily be done without having to develop each parent genotype into a network. If a certain parameter of the code for a neuron corresponds to its vertical position in the virtual space, then crossover just involves travelling down each genome, looking at each instance of this parameter and deciding by its value which of the offspring genotypes the code for that neuron belongs to.

[2] Clearly crossover based on phenotypic organisation is not biologically plausible, however this does not detract from its functional efficacy as a recombination operator.

3 Neural network problems

The vast majority of neural networks that have been employed in Evolutionary Robotics involve neurons whose output is some threshold-function of the weighted sum of their inputs. These include simple Perceptrons, multi-layer Perceptrons [3, 19], more biologically inspired continuous time neurons [17], and neurons that use different channels for excitation and inhibition [10].

One characteristic of this sort of network is that parameters need to be carefully selected to minimize the chance of a neuron's threshold lying outside the dynamic range of its inputs so that it is always either 'on' or 'off'. A balance needs to be struck between the range of genetically specifiable threshold values and the possible values that the weighted sum of the inputs may attain (given the number of input connections to a neuron and the range of genetically specifiable connection weights). The threshold range needs to be large enough so that thresholds may evolve to lie within the sort of intervals around which weighted input sums are likely to fluctuate in practice. However, it should not be so large that thresholds can easily come to lie outside the range of possible weighted input sums. If this balance is not struck properly, then evolution may have great difficulty producing a network that has any sort of reactive functionality, let alone one that displays adaptive behaviour.

Normally the designer works out beforehand, given the morphology of the network, what the genetically specifiable threshold and weight ranges should be in order to maximize the chances of networks doing 'something' rather than 'nothing'. When the connectivity of the network is put under evolutionary control, however, the number of input connections to a neuron may vary widely across the network, and the possible values that the weighted sum of these inputs may attain will vary accordingly. However the designer sets the parameters, the balance between thresholds and weights will only be optimal for one particular number of inputs, and sub-optimal for all others. While this may not prove catastrophic, it does suggest that these sorts of neural networks are perhaps not ideally suited for open-ended artificial evolution.

3.1 Connectivity-independent neurons

These problems may be avoided if we abandon the idea of using a function of the weighted *sum* of the inputs to calculate each neuron's output, and instead employ some function that is independent of the number of input connections. There are potentially many ways of achieving this and for the experiments of Section 6 the following activation function was devised. At a given point in time the output for a neuron is given by:

$$output = \begin{cases} 0 & |I_e| - |I_i| < T - 1.0 \\ 1 & |I_e| - |I_i| > T + 1.0 \\ \frac{|I_e| - |I_i| - T + 1.0}{2} & otherwise \end{cases} \tag{1}$$

where T is the threshold value of the neuron, and I_e and I_i are respectively the most active of the excitatory inputs to the neuron, and the most active of the inhibitory inputs from the neuron at that point in time. This means that at any one moment a maximum number of connections that will determine the output behaviour of each neuron is fixed at two. However, the activity of any particular input will vary with the output of its pre-synaptic neuron. So the number of different inputs which can, at different times and in different combinations, determine a neuron's output, is variable.

One can imagine circumstance in which these networks would not be functionally capable of evolving particular behaviours that would be possible using more conventional networks that operate upon weighted sums. However, they are perfectly capable of generating the complex many-faceted behaviour of section 6 and they also avoid the problems associated with parameter setting for variable connectivity outlined above. One issue this raises for future research is the exact nature of the demands placed on neuron functionality by robot control.

4 Genetic operator problems

For many who use genetic algorithms on more traditional optimization problems, the process of artificial evolution occurs as an initially random population converges upon a solution, gradually decreasing the genetic diversity until an equilibrium is reached [4]. At this point, it is assumed that no significant change will occur and the process, for all practical purposes, is finished. The situation is made more complex,however, if genotypes are allowed to grow and vary in size. This is because there is always the possibility that significant change can occur *after* the rate of genetic convergence has dropped to zero. Harvey has suggested that it is only after this initial convergence phase that the real business of open-ended artificial evolution may begin [9, 8]. Genetic operators should therefore be such that even after convergence has stabilised there is sufficient generation of new genetic material to ensure the adaptive abilities of the evolutionary process.

In the context of a converged population, the role of the crossover operator is unclear. Although it may on occasion combine useful 'building blocks' and schemas [4] from two parent genotypes in the same offspring genotype, the fact that all individuals within the population are very similar means that this will not occur often enough to play a major role. Thought must be given to whether the extra unpredictable mutation effects produced by this operator are worth the gain.

If we decide against using crossover, then we must rely on random mutation in one of its many forms[3] to maintain the genotypic variation necessary for ongoing open-ended evolution. However, the question of how to apply mutations when dealing with variable length genotypes again presents us with a dilemma. Either mutation is fixed at a particular rate per genotype or it is fixed at a particular rate per unit of genotype i.e. per gene. Both have their problems. If the mutation rate is fixed at a particular rate per gene then, as the genotype grows in length, the effective mutation rate per genotype increases until we are faced with the

---

[3] We are counting any operator that has a random effect on a single genotype, such as translocation, as a form of mutation here.

problem of the error threshold [16]. This occurs when the mutation rate is so high that fit traits and individuals are lost to mutation before they have a chance to spread through and establish themselves in the population. Thus mutation overcomes the forces of selection, and the rate at which new complexity evolves drops to zero. If the mutation rate is fixed at a particular rate per genotype then the problem of error thresholds may be avoided. However, as the length of the genotype increases, the mutation rate per gene decreases in proportion to the reciprocal. Thus the more complex the neural network encoded on the genome, the slower new complexity evolves until eventually, in practical terms, evolution again grinds to a halt.

4.1 Mutation-locking

One answer to this dilemma is to use some form of 'mutation-locking' to protect 'fit' genes from the deleterious effects of major mutations. Once it has been established that a gene contributes to the fitness of genotypes that contain it, then it will usually continue to contribute to the fitness of such genotypes from generation to generation. When this will not be true is if the fitness peak that the population occupies due to the gene is a local maximum, and later on in the run evolution finds a higher maximum elsewhere whose fitness the gene does not contribute to. This is not, however, what we observe happening in an open-ended evolutionary robotics performed on a static fitness landscape according to S.A.G.A. principles. In this case, populations are converged, and once evolution has come up with a way of producing a particular behaviour it tends to stick to it. Later on, with growth of the genotype, it may evolve to produce other behaviours as well. Thus if a gene is beneficial (i.e. plays a significant role in increasing the fitness of genotypes that contain it) then we can expect it to stay in the population indefinitely. A beneficial gene could therefore be protected from major mutations without negative consequences to the evolutionary process as a whole.

If an automatic process can be found which 'mutation-locks' beneficial genes of this type then this would offer a possible solution to the problem of applying mutation in an open-ended evolutionary robotics scenario. With the correct genotypic growth rate, the number of *un-locked* genes can be kept more or less constant, with genotypes growing only as their number of locked genes increases. Therefore a constant mutation rate per gene, which maintains the degree to which new complexity is explored, will result in an effective mutation rate per genotype that is also constant, thus avoiding the error threshold.

The experiments of Section 6 employ an automatic mutation-locking process inspired by the way in which some genes on a real chromosome are more likely to undergo mutation during reproduction than others due to their evolutionary age. This process works as follows. Associated with each gene on the genotype is a number that represents the gene's 'age'. Initially, all genes are aged zero. At each offspring event, there is a small chance per gene that it will be completely deleted from the genotype. If it is not, then its age is increased by one. There is also a small chance per offspring event that a random gene, aged zero, will be added to the genotype. With this approach, the only way in which a particular gene can survive a significant number of generations is if genotypes that contain it are consistently selected to act as parents. If genotypes that do not contain it are equally likely to be selected, then after a certain amount of time we can expect the gene to disappear from the population due to the effects of genetic drift. After a gene reaches a certain age, therefore, we can be reasonably confident that it is beneficial and mutation-lock it accordingly. This age can be arrived at empirically by running the genetic algorithm on a neutral fitness landscape for several thousand generations and observing the maximum age achieved by a gene in that time. When run on the real fitness landscape, only beneficial genes will survive for significantly longer than this maximum age.

5 Evaluation problems

The artificial evolution of control architectures for simple behaviours typically involves thousands of fitness evaluations and this can be a very time-consuming process. If these evaluations are performed on robots in the real world then they must be done in real time. If they are performed in simulation, then evolved controllers may not transfer into reality unless the simulation is so complex that all speed advantages are lost. Clearly this is a problem for Evolutionary Robotics in general, but it is of special relevance to the sort of open-ended approach under discussion here: if evolving controllers are to display not one but several simple behaviours as sub-modules of a more complex overall behaviour then the problem is exacerbated.

5.1 Minimal Simulations

Recently, Jakobi has proposed new ways of thinking about and building fast-running easy-to-design minimal simulations for the evaluation of robot controllers. This methodology is described in detail elsewhere [14, 13], but since the experiments reported in this paper make extensive use of it, we offer a brief sketch here:

1. A small *base set* of robot-environment interactions that are sufficient to underly the behaviour we want to evolve must be identified and made explicit. A simulation should then be constructed that includes a model of these interactions. Since the base set will not contain all of the robot-environment interactions that can affect evolving controllers, some features of the simulation will have a basis in reality (the *base set aspects*), and some features will derive from the simulations implementation (the *implementation aspects*).

2. Every implementation aspect of the simulation must be randomly varied from trial to trial so that controllers are unable to rely on them to perform the behaviour. In particular, *enough* variation must be included so that the only practicable evolutionary strategy is to actively ignore each implementation aspect entirely.

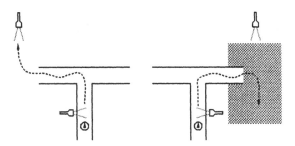

Fig. 2. The extended T-maze task: The route the robot must take for two of the eight possible combinations of light positions and floor colour.

3. Every base set aspect of the simulation must be randomly varied from trial to trial. The extent and character of this random variation must be sufficient to ensure that reliably fit controllers are able to cope with the inevitable differences between the robot-environment interaction model and reality, but not so large that they fail to evolve at all.

The power behind these ideas derives from the fact that we only have to model a sufficient number of real-world features, and these do not even have to be modelled particularly accurately. This means that such simulations can be easily constructed and made to run extremely fast. As long as the right amount of variation is included according to the methodology outlined above, controllers that evolve to be reliably fit will almost certainly transfer into reality.

6 An illustrative experiment

The preceding sections have put forward various new ideas and techniques aimed at facilitating open-ended Evolutionary Robotics. In this section we present experiments which illustrate how these techniques can be brought together to evolve complex robot behaviours. To provide a suitable challenge, the T-maze task described in [14] was extended so that only controllers that displayed a variety of reactive and non-reactive behaviours could be successful. A diagram of this extended T-maze environment is given in figure 2, and two successful paths through this environment are also shown. To achieve maximum fitness points, controllers had to guide a Khepera robot [15] down the stem of a T-maze, passing through a beam of light shining from one of the two sides chosen at random. At the T-junction, they then had to 'remember' on which side of the corridor the light had gone on and turn down the corresponding arm of the T-maze. On exiting the T-maze, they either had to steer the robot towards a second light if the floor was white, or away from it if the floor was black.

Successful completion of the full task requires the evolution of a controller capable of displaying, and mediating between, a number of different behaviours. The controller had to be capable of the following four reactive behaviours:

corridor-following, T-junction-negotiation, phototaxis and light avoidance. Additionally, the controller had to be capable of a non-reactive behaviour, namely turning the right way at the junction, which requires internal state. The controller's reaction to light also had to be relatively sophisticated. Not simply because phototaxis and light avoidance behaviours had to be employed appropriately upon exiting the maze, but also because both of these behaviours had to be inactive when the robot passed through the first light beam.

The fitness function was designed to guide the evolution of controllers through six behavioural stages in the order that they are outlined below. This progression was indeed observed each time a successful controller evolved.

1. **Corridor following:** The robot moves forward up the stem of the T-maze, culminating in hitting the top wall at the junction.
2. **Random turning:** The robot turns at the junction and exits the maze, however its turning is not correlated with the first light.
3. **Invariant reaction to second light:** Upon exiting the maze the robot either *always* moves toward the light, or *always* moves away from the light, irrespective of floor colour.
4. **Variant reaction to second light** The robot still performs only one of the two required light behaviours, but performs it appropriately with respect to the colour of the floor. For example, the robot performs phototaxis when the floor is white and simply continues along it's path when the floor is black.
5. **Appropriate reaction to second light** The robot performs both phototaxis and light avoidance appropriately with respect to the colour of the floor.
6. **Junction turning determined by stem light:** Finally, the direction the robot turns at the T-junction is always correctly determined by the stem light. It now performs the full task.

A spatially determined encoding scheme, as outlined in Section 2, was used in the experiments. Of the 8 ambient light sensors and 8 IR sensors available on a Khepera [15], only the front 6 of each were made available for use by evolving networks. All sensor values were normalized to lie within the region 0-1. In addition, a light sensor was taped to the robot's base that was thresholded to return 0 or 1 depending on the colour of the floor. The developmental space was divided up into input and output regions as shown in the diagrams of the evolved networks of figure 3. In order for a link to connect to a neuron, the distance between its genetically specified target location and the location of the nearest neuron was set at a maximum of 1/8th of the length of a side of the space. The top half of the developmental space was divided up into three horizontal strips, one for each sensor modality. The top strip and the one beneath it were each further sub-divided into 6 squares for the 6 ambient light sensors and 6 IR sensors respectively. The third strip was allocated to the floor sensor. Quarter circle shaped regions in the lower right and left hand corners of the region (each a 36th of the total area) were allocated to the right and left hand motors respectively. The values that were sent to the motors were calculated by mapping the average output of the neurons in each region from the interval 0-1 onto the interval ±10.

The neural networks described in Section 3 were used in the experiments with the output from each neuron calculated according to equation 1, Weights on the connections between neurons were in the range ±2.5 and thresholds were in the range ±1. As stated above, any input to a neuron from the sensors was normalised to lie within the range 0-1.

Genotypes were strings of a variable number of blocks of integers, or 'genes', each of which coded the parameters for a neuron. In fact, since the task was symmetrical, spatial symmetry was also imposed on evolving networks by reflection across the central vertical line of the developmental space. Therefore each block of 16 integers actually coded for a *pair* of neurons. The first of these 16 numbers corresponded to the 'age' of the gene (see below), the rest coded for a neuron's threshold, its position in the developmental space, the weights and target positions of two links *to* that neuron from other neurons and two links *from* that neuron to other neurons. All integers on the genotype, apart each gene's age, were in the range 0-99 and mapped onto the ranges of their corresponding parameters.

The genetic algorithm was an extremely simple generational model, population sized 100, which implemented an automatic mutation-locking process as described in Section 4. After every member of the population had been evaluated, two offspring each were produced asexually from the 50 fittest individuals of the population to form the next generation. Apart from the 2 offspring of the fittest individual which were identical copies, each of the other 98 had a 0.3 chance of undergoing a major 'innovative' mutation and a greater chance of undergoing a number of smaller 'tuning' mutations. The nature of each innovative mutation depended on the number of unlocked genes on the genotype. If there were less than two un-locked genes, then a random gene (aged zero) was added to the genotype. If there were more than two un-locked genes then one of these would be randomly selected and deleted from the genotype. If there were exactly two un-locked genes, then either a new gene would be added or an un-locked gene would be deleted with equal probability. This ensured that the number of un-locked genes per genotype would remain more or less constant, but that genotypes could grow as the number locked genes increased. A tuning mutation consisted of adding a random offset in the range ±5% to any of the parameters on the genome *including* those of genes that were mutation locked. The number of tuning mutations per genotype was picked from a Poisson distribution with an expected number of 2. After all offspring had been generated, the age of every gene of every individual in the population was increased by 1. Those genes that passed the age of 200 were regarded as mutation-locked and made immune from deletion.

Evolving controllers were evaluated using a minimal simulation built according to the methodology briefly sketched in Section 5. The minimal simulation of a Khepera in a T-maze described in [14, 13] was extended to incorporate a simple ambient-light sensor model.

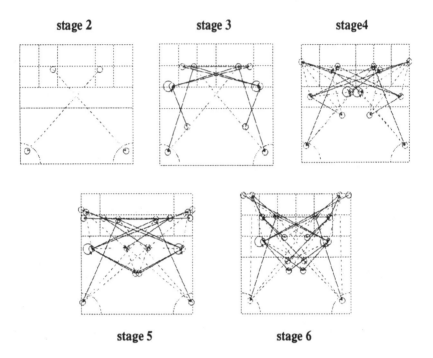

Fig. 3. Incremental evolution: The networks grow in size, complexity and capability. Each network corresponds to one of the evolutionary stages described in section 6. (Excitatory connections are shown by solid lines, inhibitory with dashed lines.)

6.1 Results

In all, several evolutionary runs were performed with various different parameter settings, and those parameters that were observed to evolve the complete behaviour in the least amount of time are reported above. A typical run, lasting 6000 generations, took around 36 hours to perform on a SPARC Ultra. This simulated over 7 years worth of real-world evolution[4].

Figure 3 shows five networks taken from the evolutionary history of a typical neural network which evolved to perform the full task. Each of these networks corresponds to a different behavioural stage, as listed above, on the route to the final behaviour. There is no network in this figure corresponding to the first stage because details of the fittest controller were only saved to file by the program every 50 generations, and by about generation 20 the fittest networks were already on stage 2. One of the advantageous aspects of using spatially determined encoding is the ease in which evolved networks may be graphically displayed.

---

[4] It must be kept in mind here that a minimal simulation is more noisy and less reliable than the real world. This means that successful behaviour would probably have evolved in less than 7 years if evolving controllers had been evaluated on the real robot.

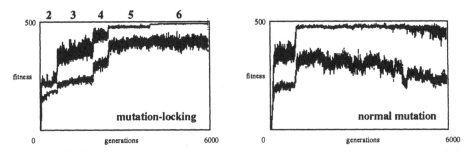

Fig. 4. Evolution with and without mutation locking.

The left-hand graph of figure 4 shows how the highest and average fitness of the population changed over time as the networks of figure 3 evolved. Note that the fitness increased in discrete jumps as the necessary new mechanism for each behavioural stage evolved and then mutation-locked. The numbers along the top of the graph correspond to each of these stages. The gap between mean and maximum fitness also stayed fairly constant, a necessary feature for ongoing open-ended evolution.

As a comparison, several runs were also performed in which the mutation-locking process was switched off and normal 'bit-wise' mutation were used instead. In order to make the comparison we needed to find suitable mutation-rate, one that was low enough to allow for the evolution of a genotype of sufficent size to encode an effective solution, yet high enough that that evolution could progress as fast as possible. In order to establish this rate we used a successful population from a previous mutation-locked run to seed a succession of GA runs with different rates of normal mutation (and no mutation locking). From these runs we were able to establish a mutation rate above which the average fitness of the population could not be maintained (i.e. an error threshold). The mutation rate for the comparision runs was then set just below this error threshold.

The right-hand graph of figure 4 shows the most successful of these comparison runs. At each offspring event, there was a chance of 0.014 per parameter of a point mutation (the addition of a random offset taken from a normal distribution with a standard deviation of 15 and a mean of 0). There was also a 0.02 chance of a random gene being added to the genotype. There was no deletion operator. As can be seen from the graph, fitness initially increased extremely quickly, but the difference in mean and maximum fitness was much greater than that observed when mutation-locking was employed. The fitness stopped increasing when controllers evolved that could perform all of the task except turning the right way at the T-junction in response to the stem light (stage 5). The complete behaviour was never observed to evolve in any of these comparison runs. Also, since there was no deletion operator, a Muller's ratchet type effect occurred as the genotype length increased inevitably. Eventually this resulted in a loss of fitness.

Transfer from simulation into reality was successful in all but one aspect. Evolved controllers satisfactorily guided the real khepera down the stem of the T-maze, turned the correct way at the T-junction, guided the khepera down the second arm of the T-maze out into the open space and, if the floor was black, satisfactorily guided the robot away from the second light. However if the floor was white on exit, the khepera curved slowly and clumsily towards the light in a less than satisfactory fashion. Upon examination it was found that the angle of acceptance of the ambient light sensors within the simulation was significantly incorrect. Once a line of compensatory code had been inserted into the program running on the Khepera, evolved controllers worked fine in reality. Controllers were subsequently successfully evolved using a corrected simulation.

7 Conclusions

This paper has shown how many of the techniques commonly used to evolve fixed-architecture neural networks for the control of robots face various problems if the size and morphology of evolving networks are allowed to vary under evolutionary control. In particular, issues pertaining to the types of encoding schemes, neural networks, genetic operators and evaluation techniques that should be used in an open-ended Evolutionary Robotics have been discussed.

This paper has also proposed some solutions to these problems and shown how these can be combined in the evolution of a complex robot behaviour. Of the new techniques put forward, some were more successful than others. Comparative studies need to be performed before anything definite can be claimed, but both the encoding scheme proposed in Section 2 and the neural networks proposed in Section 3 seemed to work well in the open-ended evolutionary experiments of Section 6. The minimal simulation used in these experiments and built according to the methodology outlined in Section 5 was not quite so successful. This was however due to careless construction rather than anything fundamentally wrong with the methodology. The performance of the mutation-locking techniques that were used in these experiments was inconclusive. As opposed to speeding up the evolutionary process, mutation-locking actually seemed to slow it down when compared to normal mutation. However, the way in which each new stage evolved and mutation-locked in turn to create the complete behaviour was encouraging. We still believe that protecting 'fit' genes from major mutation may play a significant role in an open-ended Evolutionary Robotics and are currently refining techniques by which the fitness contribution of each gene may be automatically assessed.

References

1. D. Cliff, I. Harvey, and P. Husbands. Explorations in evolutionary robotics. *Adaptive Behavior*, 2:73–110, 1993.
2. Cliff D. Ncage. neural control architecture genetic encoding. Cognitive Science Research Paper 325, Schhol of COGS, University of Sussex, 1994.

3. Floreano D. and Mondada F. Evolution of homing navigation in a real mobile robot. *IEEE Transactions on Systems, Man and Cybernetics–Part B: Cybernetics*, 26(3):396–407, 1996.

4. D. E. Goldberg. *Genetic Algorithms in Search, Optimization and Machine Learning*. Addison-Wesley, Reading, Massachusetts, 1989.

5. F. Gruau. *Neural Network Synthesis using Cellular Encdoing and the Genetic Algorithm*. PhD thesis, Ecole Normale Superieure de Lyon, 1994.

6. F. Gruau. Automatic definition of modular neural networks. *Adaptive Behavior*, 3(2):151–184, 1995.

7. F. Gruau. Cellular encoding for interactive evolutionary robotics. In P. Husbands and I. Harvey, editors, *Proceedings of the Fourth European COnference on Artificial Life*, pages 368–377, Cambridge, MA, 1997. MIT Press.

8. I. Harvey. Species adaptation genetic algorithms: the basis for a continuing saga. In F. J. Varela and P. Bourgine, editors, *Toward a Practice of Autonomous Systems: Proceedings of the First European Conference on Artificial Life*, pages 346–354, Cambridge, Massachusetts, 1992. M.I.T. Press / Bradford Books.

9. I. Harvey. *The Artificial Evolution of Adaptive Behaviour*. PhD thesis, School of Cognitive and Computing Sciences, University of Sussex, 1993.

10. I. Harvey, P. Husbands, and D. Cliff. Seeing the light: Artificial evolution, real vision. In D. Cliff, P. Husbands, J.A. Meyer, and S. Wilson, editors, *From Animals to Animats 3: Proceedings of the Third International Conference on Simulation of Adaptive Behavior*, volume 3. MIT Press/Bradford Books, 1994.

11. P. Husbands, I. Harvey, D. Cliff, and G. Miller. The use of genetic algorithms for the development of sensorimotor control systems. In P. Gaussier and J-D. Nicoud, editors, *Proceedings of From Perception to Action Conference*, pages 110–121. IEEE Computer Society Press, 1994.

12. N. Jakobi. Evolving sensorimotor control architectures in simulation for a real robot. Master's thesis, School of Cognitive and Computing Sciences, University of Sussex, 1994.

13. N. Jakobi. Half-baked, ad-hoc and noisy: Minimal simulations for evolutionary robotics. In P. Husbands and I. Harvey, editors, *Proc. 4th European Conference on Artificial Life*. M.I.T. Press, 1997.

14. N. Jakobi. Evolutionary robotics and the radical envelope of noise hypothesis. *Journal of Adaptive Behaviour*, 6(2), 1998.

15. K-Team. Khepera users manual. EPFL,Lausanne, June 1993.

16. M. Nowak and P. Schuster. Error thresholds of replication in finite populations, mutation frequencies and the onset of muller's ratchet. *Journal of Theoretical Biology, 137*, pages 375–395, 1989.

17. Beer R.D. Toward the evolution of dynamical neural networks for minimally cognitive behaviour. In P. Maes, M. Mataric, J-A Meyer, J. Pollack, and S. W. Wilson, editors, *Proceedings of the fourth international conference on simulation of adaptive behaviour*, pages 421–429, Cambridge, Mass, 1996. MIT Press.

18. Nolfi S. and Parisi D. Evolving artificial neural networks that develop in time. In F. Moran, A. Moreno, and J.J. Merelo, editors, *Advances in Artificial Life: Proceedings of the third European conferrence on Artificial Life.*, Berlin, 1995. Springer-Verlag.

19. Nolfi S. and Parisi D. Evolving non-trivial behaviours on real robots: an autonomous robot that picks up objects. In M. Gori and E. Soda, editors, *Proceedings of Fourth International Congress of the Italian Association of Artificial Intelligence*, Berlin, 1995. Springer Verlag.

Noise and the Pursuit of Complexity:
A Study in Evolutionary Robotics

Anil K Seth

Centre for Computational Neuroscience and Robotics
and School of Cognitive and Computing Sciences,
University of Sussex, Brighton, BN1 9SB, UK,
anils@cogs.susx.ac.uk

Abstract. This paper describes a new approach for promoting the evolution of relatively complex behaviours in evolutionary robotics, based on the use of noise in simulation. A 'homing navigation' behaviour is evolved (in simulation) for the Khepera mobile robot, and it is shown that high noise levels in the simulation promote the evolution of relatively complex behavioural and neural dynamics. It is also demonstrated that simulation noise can actually *accelerate* artificial evolution.

1 Introduction

The objectives of this paper are to illustrate two new ways in which the careful use of *noise* can be of benefit to evolutionary robotics; *a)* by promoting the evolution of relatively complex behaviours, and *b)* by accelerating the artificial evolution process. These objectives require that a particular approach to evolutionary robotics be adopted; namely the evolution of robot controllers in *simulation*, with successful subsequent transference of evolved controllers to the real world. The use of simulation permits the incorporation of specifiable amounts of noise[1], and the condition of transference ensures that evolutionary robotics remains faithful to real robots in the real world.

This approach, already shown to be viable (for example see Jakobi [4], Nolfi [9], Miglino et al. [8]), stands in contrast to two major alternatives. The first, evolution in real time on real robots (e.g. Floreano and Mondada [2]), does not easily allow for the explicit, quantitative specification of noise levels[2] and is also formidably time intensive. The second, evolution in simulation with significant abstraction from reality (e.g. Sims [11]), removes evolutionary robotics from the real world and therefore will not be pursued here.

In what follows, a simulation is used to evolve a 'homing navigation' behaviour; a behaviour originally evolved and investigated in the context

---

[1] Noise here consists of various aspects of a simulation incorporating degrees of randomness; these aspects are detailed in section 4. It does *not* refer to stochastic aspects of the search algorithm.

[2] Qualitative noise levels *can* be manipulated in the real world, for example by introducing flashing lights into the vicinity of the robot (see e.g. Jakobi et al. [6]).

of real-world evolution by Floreano and Mondada [2]. It is demonstrated that large amounts of simulation noise promote the evolution of robots with relatively complex behaviours and neural dynamics compared to those evolved in simulations with low noise levels. It is also demonstrated that high noise levels can accelerate the evolutionary process.

The rest of this paper is organised as follows: section 2 discusses the role of noise in simulation, and introduces some new perspectives. Sections 3 and 4 describe the experiments undertaken; briefly introducing the original real-world study by Floreano and Mondada, and then describing the simulation replication in detail. Section 5 presents some results, which are then discussed in section 6.

2 Noisy simulations

In contrast to the potential roles of noise outlined in section 1, evolutionary robotics has, to date, concentrated on the discovery that noise in simulation can help bridge the 'reality gap'. Controllers evolved in simulations with appropriate noise levels transfer to real robots and real situations (see e.g. Jakobi [5], Miglino et al. [8]).

Jakobi ([5], [4]) has formalised the use of noise for facilitating transference by distinguishing between *base set* features (those aspects of an agent-environment system that may come to play a part in the eventual behaviour) and *implementation* features (those aspects which are either simulation artefacts, or not relevant to the behaviour, or just real-world aspects that are difficult to model). The idea is then that the base-set features should be modelled noisily, so they that they will transfer to the real world, (this being base set 'robustness'), and that the implementation aspects should be made *very* unreliable, so that evolution cannot come to incorporate them in *any* viable controller, (this being base set 'exclusivity').

The use of this approach, with its integral role for noise, thus also permits the assessment of the effects of noise on the complexity of the evolved behaviour, and on the speed of evolution. Both these ideas are new to evolutionary robotics, but the former builds on previous work (Seth, [10]), in which noise is shown to promote the evolution of strategies of increasing complexity in co-evolving Iterated Prisoner's Dilemma ecologies. In these studies, the 'evolved complexity' is reflected in the strategies deployed by the agents, which in turn is reflected in the lengths of variable length genotypes. This paper is concerned with exploring the same effect in an evolutionary robotics context.

3 A 'homing navigation' experiment

The context for this study is drawn from an experiment by Floreano and Mondada 1996 [2]. Their work is briefly described below.

They demonstrate the real-world evolution of a 'homing-navigation' behaviour, using a Khepera robot [7]. This robot is equipped with an extra 'floor sensor' in addition to the usual array of 8 infra-red proximity (and ambient light level) sensors. The environment for the experiment was a 40cm by 45cm walled arena, situated in a dark room, but with a small light tower placed in one corner. This corner (denoted the 'charging area') also had black paint on the floor out to a radius of 8cm (the floor being otherwise white).

The robot had a simulated battery of 50 'actions'[3], after which it would 'die'. However, if it happened to pass over the charging area during its life, the battery would be instantaneously recharged and the robot could carry on for another 50 actions (up to an arbitrary maximum of 150).

The sensor arrangement of the robot and the (fixed) architecture of the neural net controller is shown in fig 1; with 8 input units corresponding to the 8 IR sensors, 2 input units for the ambient light sensors 2 and 6 (on the front and rear of the robot body), and one input unit each for the floor sensor and battery level. These inputs were fed through to a 5 unit internally recurrent hidden layer, which was in turn connected to a two unit motor output layer, which then set the wheel speeds. Sigmoid activation functions were employed at all layers except the input layer, which linearly scaled the sensory inputs to range from −0.5 to +0.5.

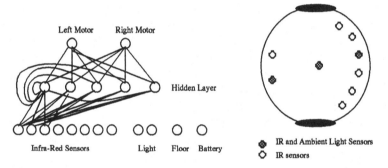

Fig. 1. *Network architecture and sensor layout. For clarity, not all network connections are shown. All input units are connected to all hidden units, and all hidden units are connected to themselves and to every other hidden unit, as well as to both motors. The central floor sensor is located beneath the robot base.*

The fitness function used by Floreano and Mondada was very simple, calculated incrementally at every step (except when the robot was directly over the charging area, where no score was awarded), and maximised by high speed and low IR input (V is the scaled average wheel speed (taken as a vector), and i the scaled activation value of the IR sensor with the highest value):

$$\theta = V(1 - i), \qquad 0 \leq V \leq 1, \ \ 0 \leq i \leq 1$$

---

[3] Each action corresponded to one update of the controlling neural network, with updates taking place about every 300ms.

There is nothing in this fitness function that explicitly specifies periodic return to the charging area, but robots that come to adopt this strategy will tend to live longer and thus accrue higher fitness than those that do not.

Floreano and Mondada performed artificial evolution in the real world, downloading candidate controllers onto real Khepera robots for each evaluation, and using a simple tournament GA to evolve the weights and thresholds of the network (which were then fixed for the duration of each individual). Over the course of ten days (200 generations) fit individuals evolved. These individuals typically explored their environment at high speed, returning to the charging area at periodic intervals, usually just in time to avoid running out of battery power. This work of Floreano and Mondada is therefore an impressive demonstration of real-world evolutionary robotics, and also provides convincing support for their primary thesis that simple fitness functions can be used to evolve relatively complex behaviours.

The thrust of the present work is different. A simulation of the Khepera robot, controller, and environment, supports the evolution of a similar 'homing-navigation' behaviour to that described above. The present hypothesis, however, is that the amount of *noise* present in the simulation strongly influences the complexity of the behaviours (and underlying neural dynamics) that evolve. It is also illustrated that noise can accelerate the artificial evolution process. A subsidiary motivation is to provide a further example (in addition to e.g. Jakobi [5], and Miglino et al. [8]) that long periods of real world evolution can be reduced to very short periods of simulated evolution, without the sacrifice of real-world efficacy.

4   The simulation

In order for the evolved behaviours to work in the real world, and to permit the incorporation of quantitatively specifiable amounts of noise, Jakobi's 'minimal simulation' methodology was followed [4]. Three important 'base set' aspects of the experiment were identified:

- the way in which the IR sensors respond, which depends on the orientation of the robot and the distance of the robot from a wall at a given angle or a corner of a given shape.
- the way in which the ambient light sensors respond to a distant light source, which depends on the orientation of the robot, and the angle of the robot to the light.
- the way in which the robot's position and orientation changes, which depend on the wheel speeds.

In order to simulate these factors, three look-up tables were employed (adapted from [4]). One held the values that IR sensors would (roughly) hold if the robot were to be positioned at 10cm from an infinitely long flat wall. This table consisted of values for all 8 IR sensors, for each of 10 robot orientations from 0 to $\pi/2$. To then calculate the actual values of the IR sensors, given walls at different distances and orientations, and for robot orientations greater than 90 degrees, the values in this table were appropriately scaled (linear scaling between 0 and 1023). If the robot happened to be in a corner, two sets of IR sensor readings were calculated, one for each of the walls involved. The maximum value for each sensor was then taken to form a composite reading. The second lookup table simply held the angles of the sensors with respect to the centre line through the robot body. The angle of each sensor to the light source could then be calculated as a function of an angle from this table, the orientation of the robot, and the angle of the robot to the light. It was then a simple matter to calculate whether or or not this angle (of sensor to light) fell within the angle of acceptance of the sensor.

The third look up table simply held values for the changes in the x and y coordinates of the robot, if it were travelling with a given speed in each of 36 different orientations. This could then be linearly scaled according to the actual speed of the robot.

The other important aspect of the simulation was, of course, that a lot of noise was employed, both during each trial and between trials (each individual was evaluated over twelve separate trials in the GA). *Intra-trial* noise was applied to the IR, light, and floor sensor values, also the robot position, orientation, and rate of orientation change during turning, and finally the position of the robot after impact with a wall. In the last case, these collisions were modelled by just randomly repositioning the robot within about 2-3cm of the wall, with a large orientation and speed change (also randomly determined). *Inter-trial* noise was applied to the angle of acceptance of the light sensors, the arena dimensions (including charging zone radius), and also the actual *levels* of IR, background IR, and ambient light noise.

In terms of Jakobi's 'minimal simulation' methodology discussed in Section 2, the intra-trial noise was designed to deliver base set robustness, and the inter-trial noise was for ensuring base set exclusivity. Thus the simulation would be expected to deliver transferable controllers. For the purpose of investigating the complexity of the evolved behaviours and neural dynamics, and the speed of evolution, two conditions were investigated; one with high levels of both inter- and intra-trial noise, and another with zero inter-trial and very low intra-trial noise. The actual levels used are given in Appendix 1.

The experiment proceeded by using a distributed GA, with a population of 100, to evolve the weights and thresholds for the controlling network as shown in fig 1. These weights and thresholds were specified as unbounded floating point numbers on a 102 allele genotype, with mutation and crossover being the only genetic operators employed[4].

---

[4] Crossover probability was set at 0.95, with a 0.03 probability of point mutation per allele; a point mutation altered the value of the allele by a random value within the range ± 0.5 (alleles were initialised within the range ± 1).

5 Simulation results

5.1 Basic performance and transference

Many evolutionary runs were performed, in both high noise and low noise conditions. With high noise levels in the simulation, evolutionary runs of about 100 generations always produced very fit individuals. These runs took about 1 hour on a single user Sun SparcUltra (143MHz) workstation, orders of magnitude faster than the real world evolution reported in [2]. Equally fit individuals evolved with low noise levels, although not nearly as rapidly as with high noise levels (this result is discussed in section 5.4). An example graph charting the progress of a 'high-noise' simulation is shown in fig 2.

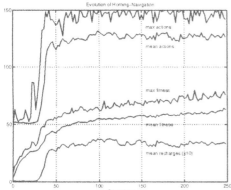

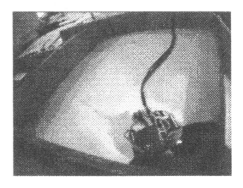

Fig. 2. *An example evolutionary run of a noisy simulation, demonstrating the evolution of fit robots in a very short time.*

Fig. 3. *A real khepera about its business, controlled by a network evolved in a very noisy simulation.*

Successful transfer to reality was consistently observed when networks from the fittest robots, evolved in noisy conditions, were downloaded onto real Kheperas (see Jakobi [4] for further examples of successful transference using the same simulation methodology). This transference was effortless, with evolved networks effectively controlling real robots without any further 'tinkering' of the evolved network or the real environment (which consisted of a cobbled together arrangement of wooden walls and a torch balanced at one corner). Fig 3 illustrates a real Khepera (powered externally, but with all processing on-board) just leaving the recharging area halfway through a demonstration. The trajectories traced by real Kheperas were very similar to those seen in simulation, except that the real robot very occasionally impacted with the arena walls. These collisions did not normally prevent the robot from recovering and reaching the charging area when necessary. Robot controllers evolved in non-noisy conditions, however, did *not* transfer effectively to the real world. The relevance of this is discussed in section 6.

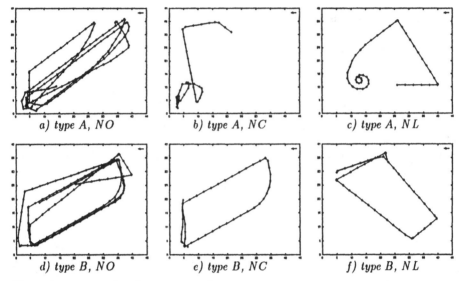

Fig. 4. *Trajectory plots (in simulation) of robots evolved under noisy (a-c) or non-noisy (d-f) conditions in either NO (a,d), NC (b,e), or NL (c,f) conditions. The charging area is located in the bottom left-hand corner of each plot, out to a radius of 8cm. The type B robots maintain a simple trajectory regardless of the environmental manipulations, but the type A robots clearly deploy more complex 'searching' and 'circling' behavioural strategies.*

The following analyses assess the contribution of high noise levels to the evolution of neural and behavioural complexity.

5.2 Behavioural analysis

A set of twelve evolved robots were analysed - six from noisy simulations (henceforth type A robots), and six from non-noisy simulations, (type B robots). All analysis took place in simulation. Initially, three environmental conditions were analysed for each robot[5]; a normal (NO) condition (with light source and charging area - this is the condition in which the robot controllers were originally evolved), a 'no charging area' (NC) condition, where the black paint is removed and the robot cannot recharge, and a 'no-light-source' (NL) condition where, although the charging area is present, the light source at the corner is removed. Low noise levels were employed in all these conditions.

Fig 4 (a-c) illustrates typical overhead trajectory plots for the robots evolved in noisy environments (type A) in the three conditions, and (d-f) illustrate the same for robots evolved in non-noisy environments (type B). In the NO condition, both A and B robots can repeatedly find the charging area (situated in

---

[5] These tests were also performed by Floreano and Mondada [2], who observed similar results to those of the 'noisy' robots in the present study. However, the conclusions drawn about the nature of the controlling network are different in the present work.

the lower left hand corner), and their trajectories are not obviously different. However in the NC and NL conditions, there are clear differences. The B robots maintain a behaviour pattern qualitatively similar to that displayed in the NO condition, but the A robots do nothing of the kind.

The B robots seem only to have evolved to move in straight lines and to turn upon encountering walls; a strategy which does indeed periodically return the robot to the charging area in predictable environments. The A robots, by contrast, are clearly affected by the presence (or absence) of the black charging area and the light source. In the NC condition, these robots navigate towards the charging area and remain in the vicinity, performing what could be described as a *searching* behaviour. In the NL condition, the robots begin, as in normal conditions, with a fairly linear trajectory, but shortly begin to *circle*, giving the impression that the robot is trying to orient to a light source.

These searching and circling behaviours were also observed in real-world Khepera behaviour, when the environment in fig 3 was manipulated in the appropriate way. All six A robots displayed similar searching and circling behaviours, and all six B robots displayed the simple behaviour (as in fig 4 (d-f)).

In a further example of how the A robots deploy more complex behaviours than B robots, one of each kind were compared in a (low noise) condition in which the walls were removed (again in simulation), with the charging area then extending in a complete circle around the light source. Fig 6 illustrates that the B robot was completely impotent in such circumstances, hinting at its reliance on IR stimulation. Only one typical run is shown, but out of 40 tests the robot reached the charging area just once. By contrast, the A robot reached the charging area 10 times out of 40, and in 4 cases returned more than once. Fig 5 illustrates a particularly impressive A robot trajectory, and although in general the robot is undeniably adversely affected by the lack of walls, the considerably greater success rate enjoyed by the A robot suggests that a greater range of environmental stimuli (not just IR) is being assimilated in the determination of its behaviour. Indeed, the A robot is clearly able to turn in the absence of a wall, and the B robot is not.

5.3 Neural analysis

In this section, it is shown that the neural dynamics of the type A robots are also more complex than those of the type B robots, and in ways commensurate with their behavioural differences.

Initially, activation plots for all 19 neurons in all three (simulation) conditions (NO, NC, and NL) for all of the 12 robots (6 A, and 6 B) were collected. Figs 9 and 10 present example plots for one robot of each type in the NO condition. These plots illustrate, (and this is true for *all* the plots), that whereas the hidden units (HUs) of the B robots react almost solely to IR stimulation, the HUs of the A robots react much more strongly to light, battery, and floor sense data. For example, at time-steps 35, 60, 95, and 125 in fig 9, hidden unit and motor output can be seen when there is *no* IR input. This is never the case in fig 10. Indeed, in the 'no-wall' condition discussed in section 4.3 (and therefore in the

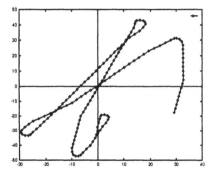

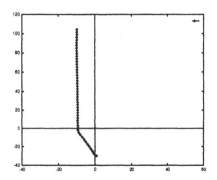

Fig. 5. *Type A robot in no-wall test; note that the edge of the graph does not represent a wall, and that the charging area is situated in a circle around the origin.*

Fig. 6. *Type B robot in no-wall test; again the edge of the graph does not represent a wall, and the charging area is a full circle around the origin.*

absence of any IR input) only the *A* robots display any significant HU activity (figs 7 and 8).

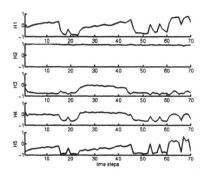

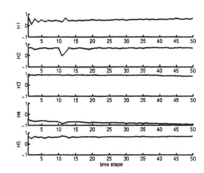

Fig. 7. *Type A robot HU activation in the no-wall test; significant levels of HU activation (only first 70 time-steps shown).*

Fig. 8. *Type B robot HU activation in the no-wall test; very little activation in any units.*

To explore these results in a non-behavioural context, short periods of high activity (spikes) were injected into six combinations of input units, with the subsequent activations of the hidden and motor units being recorded. The first two conditions consisted of IR inputs only; with either all 8 inputs active, or all except the rear two[6]. The next two conditions tested combinations of ambient light input spikes in the absence of IR input. The fifth condition injected a negative floor sensor spike (as if the robot were over the charging area); again in

---

[6] 'Non-active' units, in all conditions, were set to 0, except for floor sensor and battery units, which were set to 1 (signifying a full battery, and a robot position away from the charging area).

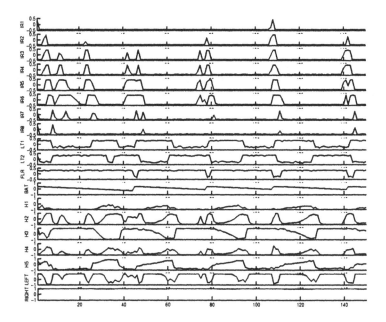

Fig. 9. *Neuron profile for type A robot (evolved in noisy conditions), tested in the NO condition. HUs (H1-H5) and motor units display activity not correlated with IR input (notably H1,H3,H5 - see time steps 35, 60, 95, and 125).*

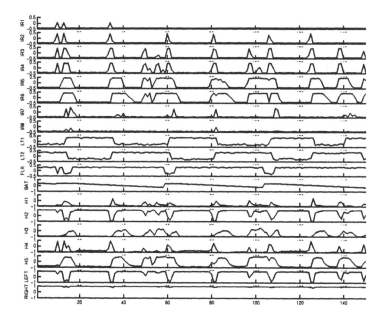

Fig. 10. *Neuron profile for type B robot (evolved in non-noisy conditions), tested in the NO condition. HUs (H1-H5) seem to be predominantly reacting to IR input.*

the absence of IR input, and the last condition injected a negative battery spike (signifying an empty battery), also in the absence of IR input.

Fig 11 presents summary data for all 12 robots over all these six conditions, in terms of the HU activity elicited. For example, the third and fourth conditions involved light input spikes, and 45 percent of the type A robot HUs responded strongly in these conditions, compared to 18 percent of the type B robot HUs. Similarly, the sixth condition tested the responses to battery sense data, and again more A HUs responded than B HUs. Thus, fig 11 makes it clear that the A robots take greater account than the B robots of the light and battery sense data. These conditions were statistically significant according to Mann-Whitney U tests (($U = 57.0; df = 6, 6; p < 0.01$), ($U = 56.5; df = 6, 6; p < 0.01$) respectively). And although the statistical test is not significant, the B robots do appear to rely more heavily on IR input than the A robots.

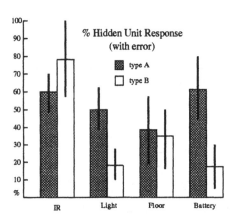

Fig. 11. *Hidden unit response patterns for both type A and type B robots. See text for details.*

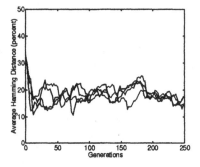

Fig. 12. *Convergence statistics for 4 runs of a noisy simulation (superimposed). Population converges rapidly to an average Hamming distance of about 20 percent.*

The results in this section are clearly suggestive of more complex evolved neural dynamics from noisy simulations than from non-noisy simulations (of course, the neural architecture itself is the same in both cases). This additional complexity is revealed primarily through type A HUs that respond to a wider range of environmental stimuli. The neural dynamics must therefore also cope with melding these multimodal inputs into a coherent motor output. These observations corroborate the behavioural data in that only the type A robots were strongly affected by manipulations of the light source and charging area. This suggests that these robots navigated to the charging zone by using the light, and could act on the basis of some internal state influenced by remaining battery level.

5.4 Noise accelerates artificial evolution

As suggested in section 5.1, it was indeed observed that high noise actually *accelerated* the evolutionary process, Figs 13 and 14 illustrate this effect, superimposing plots (of many evolutionary runs) of the number of actions taken (the lifetime) for the fittest robot of each generation, for many generations. This result is discussed in the following section.

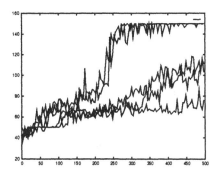

Fig. 13. *Evolution without noise. The fittest robot of each simulation reaches the maximum lifetime relatively slowly.*

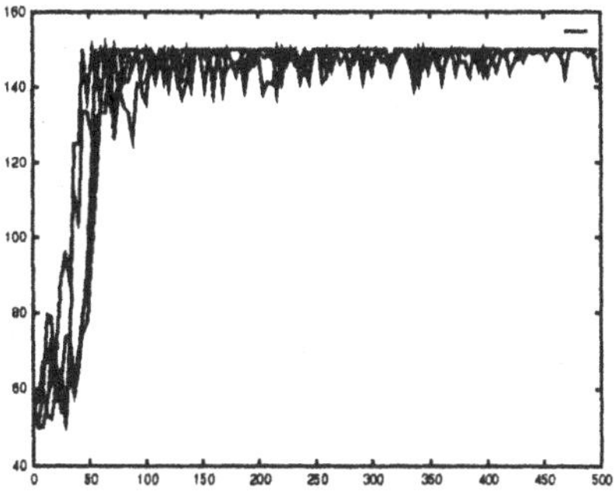

Fig. 14. *Evolution with noise. The fittest robot of each simulation reaches the maximum lifetime relatively quickly.*

6   Discussion

This paper has demonstrated how the appropriate use of noise in simulation can enhance the behavioural and neural complexity of the evolving robot controllers, and can even accelerate the evolutionary process itself. This section briefly discusses how this enhanced complexity may relate to issues of robustness, and (informally) how the effects of noise, in the present context, may be considered in terms of evolutionary dynamics.

6.1 How do these results relate to robustness?

In section 5.1 it was observed that networks evolved in non-noisy environments did *not* transfer effectively to the real world (this, however, was to be expected from Jakobi et. al [6]). One effect of noise is therefore to endow the evolving networks with real-world robustness. Indeed, one way to understand the results presented here is that robustness is being achieved through evolution exploiting the properties of a qualitatively different, more complex, behaviour. This evolutionary strategy is possible because the fitness function employed is sufficiently simple and non-specific, so to allow different behaviours to achieve similar fitness scores[7] (as noted in section 5). Future work will further address this issue of robustness, by investigating behavioural differences engendered by two different levels of noise, with *both* levels being sufficiently high to support transference to the real world.

6.2 Some informal speculations in evolutionary dynamics

Hinton and Nowlan [3] show how *learning* can influence the course of evolution, through the Baldwin effect [1]. Specifically, they show that sharp peaks in a fitness landscape can be considerably smoothed if lifetime learning is allowed. As an informal speculation, it may be that simulation noise is playing a similar role in the present experiments, in 'smoothing out' any steep valleys that may lie between maxima representing type B and type A behaviours[8]. This assumes that the population travels across a landscape as a relatively converged mass (fig 12 illustrates that the populations in the present experiments do converge quickly).

To continue this informal discussion, noise selectively applied to only *parts* of a simulated agent-environment system, could smooth some parts of a fitness landscape more than others, and perhaps entirely eliminate certain local maxima. In the present example, sufficiently high noise levels ensure that the simple type B behaviour is not viable, and the corresponding maxima would no longer be present in the landscape. Evolution could then proceed directly towards the maximum representing the type A behaviour. Finally, and informally once again, smoothing a fitness landscape with noise may reduce the total number of maxima present in the landscape, thus allowing evolution to proceed with greater speed to its ultimate maximum. This may help to explain the results presented in section 5.4.

Appendix 1

The noise levels used in the simulations are given overleaf:

---

[7] It is, of course, possible that a more specific fitness function would allow a type A behaviour to evolve in a non-noisy environment; however this paper does *not* claim that noise can permit the evolution of behaviours that would be impossible to evolve without noise.

[8] The learning algorithm employed by Hinton and Nowlan is simply random search, which could be informally construed as 'fitness evaluation noise'.

	noisy levels	non-noisy levels
IR (and background IR)	±50(±10)	±10(0)
light (and floor sensor)	±50	±5
robot position	±0.1cm	0
robot orientation	±0.02rad	0
turning noise	±0.2rads	0
friction	±3cm	0
arena size	±5cm	0
charge radius	±1cm	0
light angle of acceptance	±0.25rad	0

Acknowledgements

Thanks to Nick Jakobi and Adrian Thompson for help with the robots, and also to Matt Quinn, Andy Philippides, Tom Smith, Phil Husbands, Inman Harvey, and Hilary Buxton for good advice. Financial support was provided by the EPSRC award no. 96308700.

References

1. J.M. Baldwin. A new factor in evolution. *American Naturalist*, 30:441–451, 1896.
2. D. Floreano and F. Mondada. Evolution of homing navigation in a real mobile robot. *IEEE transactions on systems, man, and cybernetics: part B; cybernetics*, 26(3):396–407, 1996.
3. G.E. Hinton and S.J. Nowlan. How learning can guide evolution. *Complex Systems*, 1:495–502, 1987.
4. N. Jakobi. Evolutionary robotics and the radical envelope of noise hypothesis. *Journal of Adaptive Behaviour*, 6(2), 1997. (forthcoming).
5. N. Jakobi. Half-baked, ad-hoc and noisy: Minimal simulations for evolutionary robotics. In P. Husbands and I. Harvey, editors, *Proceedings of the 4th European Conference on Artificial Life*, pages 348–357. MIT Press, 1997.
6. N. Jakobi, P. Husbands, and I. Harvey. Noise and the reality gap: the use of simulation in evolutionary robotics. In F. Moran, A. Moreno, J. Merelo, and P. Chacon, editors, *Advances in Artificial Life: Proc. 3rd European Conference on Artificial Life*. Springer-Verlag, 1995.
7. K-Team. Khepera: the user's manual. Technical report, LAMI-EPFL, 1993.
8. O. Miglino, H.H. Lund, and S. Nolfi. Evolving mobile robots in simulated and real environments. *Artificial Life*, 2(4):417–434, 1996.
9. S. Nolfi. Evolving non-trivial behaviours on real robots: a garbage collecting robot. Technical Report 96-04, Institute of psychology, NRC, Rome, Italy, 1996.
10. A.K. Seth. Interaction, uncertainty, and the evolution of complexity. In P. Husbands and I. Harvey, editors, *Proceedings of the 4th European Conference on Artificial Life*, pages 521–530. MIT Press, 1997.
11. K. Sims. Evolving 3d morphology and behaviour by competition. In *Proceedings of Alife IV*, pages 28–39. MIT Press, Cambridge, Mass., 1994.

Hardware Solutions for Evolutionary Robotics

Dario Floreano[1] and Francesco Mondada[1,2]

[1] Laboratory of Microcomputing
Swiss Federal Institute of Technology, CH-1015 Lausanne
[2] K-Team SA, CH-1028 Préverenges, Switzerland

Abstract. Evolutionary robotics— as other adaptive methods, such as reinforcement learning and learning classifier systems—can take considerable time and resources which require a careful evaluation of the hardware tools and methodologies employed. We outline a set of hardware solutions and working methodologies that can be used for successfully implementing and extending the evolutionary approach to complex environments, robots, and real-world applications. The issues discussed include the integration of simulation and real robots, design issues of evolvable robots, hardware requirements for incremental evolution, and hardware and software tools for monitoring and analysis.

1 Introduction

Evolutionary techniques applied to robot control can generate efficient, smart, and creative solutions which match the constraints imposed by the environment and the selection criterion. The power, flexibility, and generality of artificial evolution has often been exploited both for finding engineering solutions to difficult control problems of autonomous robots and for gaining new insights into the biological mechanisms of adaptive behavior (see, e.g., [25, 5, 23, 29, 19, 3, 21, 14], for thematic collections of several relevant papers). However, evolutionary robotics— as other adaptive methods, such as reinforcement learning and learning classifier systems—can take considerable time and resources [24] which require a careful evaluation of the hardware tools and methodologies employed.

In this contribution, we shall outline a set of hardware methodologies jointly developed and tested at our laboratory and at K-Team SA for reliable and controlled evolutionary experiments with mobile robots. The design issues and considerations described below can be used as a guideline for assessing the requirements and feasibility of a planned experiment in evolutionary robotics, but can also be generalized to experiments involving other types of learning methods. Although most of the solutions proposed in this paper are applicable to mobile robots only, the underlying methodological issues can be extended also to other robotic platforms equipped with adaptive control systems.

The paper is articulated around four major issues. In the next section we shall discuss a way to combine the advantages of simulations with those of real robots, introducing the concept of miniature mobile robots. We shall then discuss basic design principles and methodologies of robots that can be used for artificial evolution. Capitalizing on these principles, we shall also discuss hardware solutions which can support incremental evolution, namely modularity and cross-platform compatibility. Finally, we shall address the issue of monitoring and analysis of evolved robotic controllers.

2 Between Simulation and Real World

Evolutionary robotics has its roots in simulations. The first explorations in evolution of autonomous sensorimotor agents date back to the end of the Eighties,[3] when all studies were still carried out in computer simulations. A few years later, the appearance of more robust, flexible, and user-friendly robots, and a general awareness of the limitations of simulation methods [2], created a strong motivation for the first physical implementations of evolutionary robots [9, 17].

Despite the importance of keeping in mind the hard constraints of operating with physical robots, simulations still play an important role in evolutionary robotics for at least two sets of reasons:

- **Why simulations: The practical reason.** Researchers who actively contribute to this field have very different backgrounds, including computer science, robotics, psychology, biology, and philosophy. Only few of them have the technical skills and/or local support to build a robot and program it. Most of the commercial platforms available on the market have been developed by robot experts for robot experts, resort to very specific programming languages, and thus are not accessible to outsiders. This situation is currently improving, but a majority of robots still needs specific know-how. Finally, in universities, where software writing is considered "without costs", a real robot seems often too expensive in comparison to simulations.
- **Why simulations: The strategic reason.** Evolution of complex behaviors from scratch on a physical robot, even where technically feasible, would require an adaptive time which is too long to be practically exploited for industrial applications. Simulations can be of great help when properly integrated with tests on the physical robot. Current simulative methods have progressed from unrealistic grid-worlds to new methodologies that guarantee an acceptable transfer to the target robot under well-understood constraints [31, 26, 20] and can speed up evolutionary search by several orders of magnitude. One can develop initial behavioral skills in simulation and then continue on the physical robot, or combine evaluations on simulated and physical robots, or even build computationally efficient simulations that allow a good transfer to the real robot. Furthermore, we think that simulation is an important first step within an incremental evolutionary approach, which will be discussed below in section 4.

The miniature mobile robot Khepera has been designed as a research tool to fill the gap and allow a smooth transition between simulations and the real

---

[3] But appeared in the scientific press only a couple of years later [6, 32, 8]

Fig. 1. The miniature mobile robot Khepera beside a ruler (in cm). Black pucks around the body are active infrared sensors. Rechargeable batteries are sandwiched between the motor base and the top structure; the latter hosts the microcontroller (black square), EPROM, RAM, and communication ports.

world. Initially conceived in 1991 at our laboratory by E. Franzi, A. Guignard, and F. Mondada, Khepera is currently employed in several hundred laboratories around the world for research in traditional and new-wave robotics.[4] If on the one hand its simplicity and interface make the Khepera as easy to work with as a simulation environment, on the other hand this robot cannot reach the full complexity of real-world applications. However, being a physical robot operating in a physical environment, it displays most of the characteristics of robots used for real-world applications. Therefore, it is well-suited for initial developments when one still wishes to run its own evolutionary models on the computer, but also wants to interface the program with a robot in a plug-and-play fashion in order to include real-world features in the evaluation of the genetic strings.

One of these features is *real-time operation.* In robot simulations, time is often a commodity that can be ignored or easily managed. All processes are synchronized, sub-routines are sequentially executed, the speed of sensorimotor loops is dictated by the number of intervening computations and by the power of the workstation CPU, and energy resources are not an issue. On the other hand, time constraints are ubiquitous in real robots and in biological organisms and, in the latter case, are often exploited by natural selection (for example, in several predator-prey scenarios). At the same time, several researchers have acknowledged the importance of evolving neural networks with temporal dynamics [1, 10, 18]. Khepera is a robot that can be attached to the serial port of a workstation and handled in the same way in which one would handle a simulator, but it also offers all time constraints of real-world robots. Depending on the working modality chosen by the user (see next section), the behavior of the robot is affected by the time taken by routine computation, the length of

---

[4] Here we will focus mainly on general design principles suitable for evolutionary robotics; a technical description of the Khepera robot can be found elsewhere [28].

messages exchanged by individual sub-components, and the type of communication between the robot and the workstation (if any). Here evolution of neural controllers with time dynamics has two advantages. Neurocontrollers can either adapt their own internal dynamics to those imposed by the experimenter on the robot (such as fixed action duration), or can exploit asynchrony and physical time delays to better cope with the environment. Furthermore, if controllers are evolved on a real robot, there is no need to include extra components in the fitness function in order to penalize complex architectures: if fast-thinking is important, sleek architectures and simple smart mechanisms will have higher reproduction chances over architectures that require heavy computations and complex memory handling.

Another feature that Khepera and real-world robots share is sensor and motor imprecisions. This well-known difference between simulations and real robots is of great importance in evolutionary robotics because artificial evolution often generates controllers that rely on the sensorimotor characteristics of the agent [9] more than control systems designed according to human logic. Interfacing one's own evolutionary software with a real robot not only generates more robust controllers, but can also qualitatively affect the type of controllers which are evolved.

3 Evolvable Robots

Building mobile platforms suitable for intensive learning mechanisms, such as artificial evolution, requires specific solutions. Although most of those outlined below have been incorporated in various stages of the realization of the Khepera robot, they are general enough to be used as guidelines for building other platforms.

3.1 Miniaturization

During the initial generations of an evolutionary run, most of the individuals display behaviors which might damage the robot shell, such as collisions against obstacles, improper handling of manipulators, pushing walls at sustained speed, etc. Whereas large robots have little chance of remaining operative for more than a few generations under these conditions, fundamental laws of physics endow miniature robots with higher mechanical robustness. In order to intuitively understand this feature, compare a robot of 50 mm in diameter crashing against a wall at 50 mm/s to a robot of 1 m in diameter crashing against the same wall at 1 m/s. The miniature robot will resist the collision, the other robot will probably report serious damages. However, it is important to choose the appropriate level of miniaturization. Although 1 cubic inch autonomous mobile robots are technically feasible [30], it would be advisable to choose a level of miniaturization where there is still a sufficient flexibility at the level of the development tools available, the components used, the performances achieved and the effort required by the development. Currently, the minimum attainable size for an autonomous learning robot which can be assembled with off-the-shelf components is around 2-3 cubic inches.

3.2 Interface granularity

Artificial evolution should be allowed to access sensors, motors, and other features of the robot at several levels, including the lowest-possible level. By restricting user's access to high-level commands and pre-processed sensory information, one certainly reduces the degrees of freedom of the controller, but might also hamper the adaptive power of the evolved solution. Evolution should be allowed to manipulate and combine fine details of the sensorimotor interface in the the most suitable way for the characteristics of the environment where the robot operates.

Think, for example, about the speed of the wheels. Some of the possible command levels are: go forward (speed and action duration are fixed); set speeds of wheels (action duration is fixed); set speeds and duration. One can go even further down, by-passing any PID algorithm to control motor speed. High-level commands are better suited for well-known and noise-free environments, whereas low-level commands give more adaptive power in unknown situations with variable levels of noise. These are typical situations where one wishes to use artificial evolution, rather than pre-assembling a set of macro-behaviors.

A similar reasoning goes for the sensory input. Pre-processed information is an efficient solution if one knows the features of the environment where the robot will operate. In other cases, raw sensory data, opportunistically exploited by evolutionary mechanisms, are likely to generate more robust controllers. For example, the vision module K213 [27] for the Khepera, a linear array of 64 photoreceptors spanning a 36° visual field, offers both low-level and high-level access to the data. Low-level access provides the gray-level image refreshed at a variable rate, whereas high-level functionalities allow on-board extraction of the pixel with the minimal or maximal activity, reading of only a part of the image, trasmission of thresholded values, etc. Clearly, pre-processed information might reduce the evolutionary search space and amount of computation required by the evolutionary controller, but it is necessary to ponder accurately whether this might *a priori* exclude the emergence of interesting solutions.

In contrast to robots employed for pre-programmed actions in well-known environments, the sensorimotor interface of evolvable robots should be organized in layers offering different levels of granularity, all directly accessible in parallel by the evolutionary engine.

3.3 Interface quality

Scientists and engineers develop over the years a preference for a set of standard computing and visualization tools. Just as biologists prefer to give public presentations with slides whereas computer scientists rather prefer transparencies, different research communities have different ways of working with a computer.

Getting used to a new software or to a new machine is often perceived as a nuisance and a waste of time.

Robots developed for an interdisciplinary research enterprise, such as artificial evolution, should be easily interfaceable to different platforms, include libraries for major language compilers, and be easily integrated with popular scientific packages. Plug-and-play feel is very important in this cross-disciplinary field because it encourages an increasing number of people to move from simulations to real robots. Flexibility, transportability, and an open architecture are other basic qualities of the interface of a robot intended for several purposes and user types.

3.4 Computation management

Artificial evolution requires three basic features: sustained electrical power for long periods, data storage for analysis, and some computational power. In order to meet these requirements, which are also shared by other learning mechanisms, the Khepera robot has been designed so that it can be attached to any computer through a serial connection and rotating contacts (figure 2). The serial connection provides electrical power and supports fast data communication (up to 57kBaud); it is simpler to implement and more reliable than infrared and/or radio links. This setup allows the user to run the evolutionary algorithm and all the control routines on the computer CPU which reads sensor activity and sends motor commands several times a second through the cable. All data concerning several generations can be conveniently stored on the computer hard disk for later analysis.

Real-time sensorimotor communication between the robot and the workstation, however, shows its limitations when the amount of data to be transferred is relevant, such as with vision, and when the environment requires fast responses (dynamic environments). Furthermore, the user might wish to test or further train an evolved controller on large environments without a cable connection. Autonomous robots thus must have more computational power and memory resources on-board than other types of robots, such as tele-operated robots or pre-programmed machines. If sufficient memory and computational power is available on-board, the evolved control system can be cross-compiled for the processor on the robot and downloaded.

Alternatively, one can run the control system on the robot itself and the evolutionary algorithm on the computer. This latter hybrid solution requires only one data transmission per individual for sending the artificial chromosome from the computer to the robot (and optionally reading the fitness value returned by the previously tested individual). We have used this method for predator-prey co-evolutionary experiments with two Khepera robots, where any delay in visual processing and/or motor reaction due to transmission time would be readily exploited by the opponent [11].

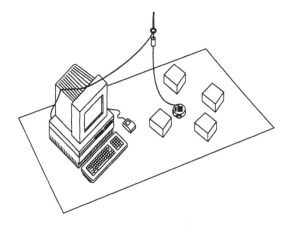

Fig. 2. The robot communicates with a computer via a thin suspended serial cable which also supplies electrical power. Rotating contacts prevent cable twisting and ensure noise-free communication. This solution can be used to control the robot from the computer, download the evolved controllers on the robot processor, or use a hybrid approach (see text).

4 Incremental Evolution

The future and applicability of evolutionary robotics heavily relies on methods of incremental evolution. Incremental evolution consists in gradually evolving a controller on a series of tasks of increasing complexity. There are several reasons why an incremental approach is important in evolutionary robotics.

Attempting to evolve a complex task from a limited collection of randomly created strings is difficult because none of the individuals in the initial generation might display competences which can be credited by the fitness function. Various approaches have been suggested, such as gradually increasing the complexity of the environment and/or modifying the fitness function during evolution [7, 17, 13]. In some circumstances, changing the environment is equivalent to modifying the morphology of the robot. For example, instead of attempting to evolve from scratch a complex behavior based on vision and appropriate control of a gripper module, it can be more fruitful to proceed gradually from a stripped-down version of the robot and add new components at later stages [12]. Incremental solutions that can cope with extendable hardware architectures are well-suited for open-ended evolution where the task under consideration can change over time in unpredictable ways [16] depending on current requirements.

Another reason for pursuing an incremental approach is scalability to robots suitable for real-world applications. Such robots are usually larger and more fragile than miniature robots suitable for research. We think that a viable solution consists in incrementally evolving behavioral competences starting from simulations, then gradually move to simple robots, and eventually continue on

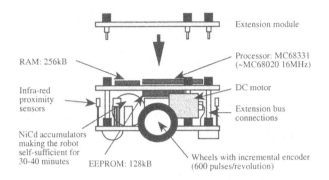

RAM: 256kB

Infra-red
proximity
sensors

NiCd accumulators
making the robot
self-sufficient for
30-40 minutes   EEPROM: 128kB

Extension module

Processor: MC68331
(~MC68020 16MHz)

DC motor

Extension bus
connections

Wheels with incremental encoder
(600 pulses/revolution)

Fig. 3. Khepera robot structure and extension possibilities.

more complex robots. Cross-platform evolution is a form of incremental evolution where previously evolved blocks are gradually adapted, combined, and extended to accomodate new morphologies, sensorimotor interfaces, and changing task requirements.

Incremental evolution requires novel strategies capable of coping with variable fitness landscapes, variable-length genotypes, ontogenetic adaptation mechanisms, variable genotype-phenotype mappings, and modularity. Interesting methods tackling some of these issues have already been suggested [15], but much work remains to be done in the years ahead. In the next two sections we will address some hardware solutions that can support investigations in incremental evolution, namely modularity and cross-platform compatibility.

4.1 Modularity

Hardware modularity enables different possible configurations and experiments using the same basic components. It also means the possibility of adding extensions and, globally, cheaper equipment. Software modularity means flexibility and possibilities for extensions as new modules modules become available on the robot.

For example, at the hardware level the Khepera robot (figure 3) has an extension bus that makes it possible to add turrets on the top of the basic configuration, depending on the needs of the experiments to be done. This modularity is based on a parallel and a serial bus. The parallel bus can be used for simple extensions directly under control of the main Khepera processor. The serial bus implements a local network for inter-processor communication. Using this second bus, other processors can be connected to the main one in order to build a multi-processor structure centred on the Khepera main processor. This kind of structure has the advantage that one can employ additional computational devices on extension modules, thus keeping the main processor free for global management of the robot behavior.

Fig. 4. Some modules of the Khepera robot. From left: basic platform with motors and batteries, microprocessor platform, linear vision, gripper, CCD camera, general input-output, and infrared communication.

Some of the hardware modules currently co-developed in the lab and at K-Team SA for the Khepera robot include a gripper with two degrees of freedom (elevation and grasping), a linear vision module (64 photoreceptors with automatic light sensitivity adjustement), a black-and-white or color CCD camera, a general input-output module where one can add several new sensors and actuators, a miniature GPS system for localisation, a radio link module, and an infrared communication module (figure 4).

At the software level, modularity is needed to support the multi-processor structure of the robot. It consists of a flexible protocol that recognizes all added extension modules when the robot is powered, informing the main processor about all functionalities available in each extension as well as the procedures for activating these functionalities. The BIOS of the Khepera, which includes all basic procedures for robot management, is also based on a modular structure. Motors, sensors, and timing functions are grouped into distinct modules to simplify management of the robot and improve software robustness. The main software also supports remote control and down-loading of specific applications through the serial cable. Such software structure simplifies the task of the user, who can easily add her own software to the management modules already implemented on Khepera.

4.2 Cross-platform compatibility

Cross-platform compatibility in mobile robotics is a very unusual feature, especially between robots of very different sizes. Software compatibility alone is not sufficient to guarantee a good transfer of algorithms between different robots.

Fig. 5. The Koala robot, despite its look and size (31 x 32 x 18 cm), has operating features similar to those of the Khepera robot.

Mechanical structure, sensor caracteristics, and many other aspects of the physical platform also play an important role in the feasibility of the transfer.

Koala (figure 5) is a medium-size very performant mobile robot that has been designed to support transfers from the Khepera robot. Despite its better performance, size, shape, and look, the Koala robot is very similar to the Khepera in many essential aspects. The six wheels of Koala are driven by two motors, as for Khepera, each controlling one side of the platform. The central wheels on the two sides are lower than the others, so that rotation of the whole platform is very similar to that of the Khepera. The proximity sensors of the Koala are based on the same concept used for those of the Khepera, but the measurement range is scaled up to the larger size of the Koala. Also, the number of sensors has been changed from 8 on the Khepera to 16 on the Koala. The Khepera hardware modularity described above has also been both supported and scaled up in the Koala. In addition to the serial extension bus of the Khepera, the Koala is equipped with a fast inter-processor parallel communication bus to support larger amounts of data. At the software level, both robots are compatible, having the same low-level BIOS software and the same comunication facilities.

All these caracteristics allow a smooth transfer from the Khepera to the Koala, just like from simulations to the Khepera. The development and monitoring tools are the same for both robots and the code itself, even when downloaded (see section 3.4 above), is very similar. Preliminary screening of the evolving populations can start on simulations, than gradually move to the Khepera, and eventually continue on the Koala. In such a framework, the incremental approach suggested by Harvey [15] seems very interesting because, after an initial stage, evolving populations are almost always converged. That means that the

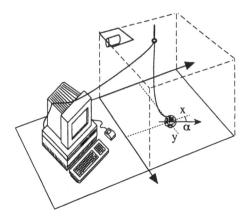

Fig. 6. The laser positioning device is composed of two elements; a portable laser emitting device on the ceiling of the room and an additional module plugged on the top of the robot which measures parameters of the laser beam and computes the exact robot position. This information is used only for neuroethological analysis.

number of maladaptive individuals tested on the physical robots is very limited with obvious benefits for the hardware platform.

5 Monitoring and Analysis

Artificial evolution requires tools for monitoring the evolutionary process and analysing the evolved controllers. In engineering applications monitoring and analysis are necessary for ensuring the stability of the learning process and of the evolved machine; in artificial life, they are required to understand the course of evolution and to establish relationships between evolved solutions and environmental constraints.

When artificial evolution is run in simulation, all the variables are readily available and analysis is technically straightforward. When it comes to evolving real robots, experimental results are often displayed as hand-drawn trajectories, photographs of the environment, or—at best—statistics on discrete events, such as number of collisions, of collected objects, etc. These methods are not sufficient to understand the tightly coupled dynamics between the evolved controllers and the environment. In contrast to disembodied machine learning systems (e.g., neural networks trained on pre-assembled data sets), evolutionary controllers develop mechanisms that depend on the type of interactions with the environment; the input distribution is determined both by the environment features and by the motor actions of the agent, which—in turn—depend on the sensory input [32]. This implies that one cannot hope to understand an evolved controller by isolating it from the environment and looking at its structure.

Understanding the course of artificial evolution and the functioning of evolved

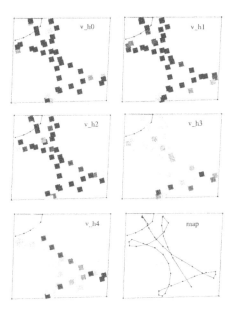

Fig. 7. Neural correlates of behavior. The evolved robot navigates in an environment and periodically returns to a recharging station. The controller is a neural network with five hidden neurons and fully recurrent connections. Each box plots the activity of one neuron every 100 ms while the evolved robot freely moves in the environment. Darker squares mean higher node activations. The robot starts in the lower portion of the arena. The recharging area is visible in the top left corner of each box. The bottom-right window plots only the trajectory. From [10]; ©1996 IEEE.

mechanisms requires a method to measure and correlate controller states with behavioral patterns. This implies two efforts. On the hardware side, the robotic setup must be designed so that physical displacements of the robot can be recorded on a fine time scale. On the software side, it should be possible to visualize, analyze, and correlate all the data (robot actions and controller states) at several levels. This procedure is similar to that employed in neuroethology where measurable sensory stimuli and motor actions are correlated with neural activity *in vivo* while the organism freely interacts with a controlled environment.

Data acquisition and analysis can be achieved in different ways. For example, a solution developed at the University of Edinburgh consists in placing LEDs on the robot and using a camera connected to a frame grabber which provides position data in real time [22]. However, this set up does not allow neuroethological analysis, unless also controller data are integrated with the position acquisition software. At our laboratory we have tested another approach, using the design principles described above (figure 6). A device emitting laser beams at predefined angles and frequencies was positioned in the environment and the Khepera was equipped with an additional turret capabable of detecting laser beams and computing in real-time the robot's displacements. This computation was carried out on a processor placed on the additional turret which did not interfere with the main controller on the robot. Every 100 ms robot position and controller variables where sent to a programmable acquisition software [4] which instantaneously processed and visualized data while the robot freely interacted with the environment (figure 7).

6 Conclusion

It has been claimed elsewhere that evolutionary robotics might be limited to simple experiments in highly controlled environments and might be applicable only to few robotic platforms [24]. In this paper we have outlined a set of hardware solutions and working methodologies that can be used for successfully extending the evolutionary approach to complex environments, robots, and real-world applications. The principles described in this paper are simple and general enough to be implemented on a variety of different robotic setups, and have all been exploited during various realization phases of the robots developed at our laboratory.

Artificial evolution remains a very powerful and general technique for automatically developing autonomous robots expected to operate in partially unknown environments and to investigate mechanisms of biological adaptive behavior. The time required to generate interesting solutions (which is comparable or shorter than the time required by other learning techniques, such as reinforcement learning) does not imply that the approach is not viable; it simply means that one should use it only for problems where traditional AI methods fail to produce interesting and useful controllers.

Acknowledgements

The authors wish to thank Edo Franzi and André Guignard for important contributions in the design of the hardware components described above. Khepera and Koala are trademarks of K-Team SA.

References

1. R. D. Beer and J. C. Gallagher. Evolving dynamical neural networks for adaptive behavior. *Adaptive Behavior*, 1:91–122, 1992.
2. R. A. Brooks. Artificial Life and real robots. In F. J. Varela and P. Bourgine, editors, *Toward a practice of autonomous systems: Proceedings of the First European Conference on Artificial Life*. The MIT Press/Bradford Books, Cambridge, MA, 1992.
3. R. A. Brooks and P. Maes, editors. *Artificial Life IV. Proceedings of the Fourth International Conference on Artificial Life*, Cambridge, MA, 1994. MIT Press.

4. Y. Cheneval. Packlib, an interactive environment to develop modular software for data processing. In J. Mira and F. Sandoval, editors, *From Natural to Artificial Neural Computation, IWANN-95*, pages 673–682, Malaga, 1995. Springer Verlag.

5. D. Cliff, P. Husbands, J. A. Meyer, and S. W. Wilson, editors. *From Animals to Animats: Proceedings of the Third International Conference on Simulation of Adaptive Behavior*, Cambridge, MA, 1994. MIT Press/Bradford Books.

6. D. T. Cliff. Computational neuroethology: a provisional manifesto. In J. A. Meyer and S. W. Wilson, editors, *From Animals to Animats: Proceedings of the First International Conference on Simulation of Adaptive Behavior*. MIT Press-Bradford Books, Cambridge, MA, 1991.

7. D. Floreano. Emergence of Home-Based Foraging Strategies in Ecosystems of Neural Networks. In J. Meyer, H. L. Roitblat, and S. W. Wilson, editors, *From Animals to Animats II: Proceedings of the Second International Conference on Simulation of Adaptive Behavior*. MIT Press-Bradford Books, Cambridge, MA, 1993.

8. D. Floreano, O. Miglino, and D. Parisi. Emergent complex behaviours in ecosystems of neural networks. In E. Caianiello, editor, *Parallel Architectures and Neural Networks*. World Scientific Press, Singapore, 1991.

9. D. Floreano and F. Mondada. Automatic Creation of an Autonomous Agent: Genetic Evolution of a Neural-Network Driven Robot. In D. Cliff, P. Husbands, J. Meyer, and S. W. Wilson, editors, *From Animals to Animats III: Proceedings of the Third International Conference on Simulation of Adaptive Behavior*, pages 402–410. MIT Press-Bradford Books, Cambridge, MA, 1994.

10. D. Floreano and F. Mondada. Evolution of homing navigation in a real mobile robot. *IEEE Transactions on Systems, Man, and Cybernetics-Part B*, 26:396–407, 1996.

11. D. Floreano, S. Nolfi, and F. Mondada. Co-evolutionary Robotics: From Theory to Practice. In preparation, 1997.

12. D. Floreano and J. I. Urzelai. Evolution and learning in autonomous robots. In D. Mange and M. Tomassini, editors, *Bio-Inspired Computing Systems*. PPUR, Lausanne, 1998.

13. F. Gomez and R. Miikkulainem. Incremental evolution of complex general behavior. *Adaptive Behavior*, 5:317–342, 1997.

14. T. Gomi, editor. *Evolutionary Robotics. From intelligent robots to artificial life*. AAI Books, Kanata, Canada, 1997.

15. I. Harvey. Species Adaptation Genetic Algorithms: A basis for a continuing SAGA. In F. J. Varela and P. Bourgine, editors, *Toward a Practice of Autonomous Systems: Proceedings of the First European Conference on Artificial Life*, pages 346–354. MIT Press-Bradford Books, Cambridge, MA, 1992.

16. I. Harvey. Artificial evolution for real problems. In T. Gomi, editor, *Evolutionary Robotics*, pages 187–220. AAI Books, Ontario, Canada, 1997.

17. I. Harvey, P. Husbands, and D. Cliff. Seeing The Light: Artificial Evolution, Real Vision. In D. Cliff, P. Husbands, J. Meyer, and S. W. Wilson, editors, *From Animals to Animats III: Proceedings of the Third International Conference on Simulation of Adaptive Behavior*. MIT Press-Bradford Books, Cambridge, MA, 1994.

18. I. Harvey, P. Husbands, D. Cliff, A. Thompson, and N. Jakobi. Evolutionary Robotics: The Sussex Approach. *Robotics and Autonomous Systems*, 20:205–224, 1997.

19. P. Husbands and I. Harvey, editors. *Proceedings of the Fourth European Conference on Artificial Life*, Cambridge, MA, 1997. MIT Press/Bradford Books.

20. N. Jakobi. Half-baked, ad-hoc and noisy: Minimal simulations for evolutionary robotics. In P. Husbands and I. Harvey, editors, *Proceedings of the 4th European Conference on Artificial Life*, Cambridge, MA, 1997. MIT Press.

21. C. G. Langton and K. Shimohara, editors. *Artificial Life V. Proceedings of the Fifth International Conference on Artificial Life*, Cambridge, MA, 1997. MIT Press.

22. H. H. Lund, E. de Ves Cuenca, and J. Hallam. A Simple Real-Time Mobile Robot Tracking System. Technical Report 41, Dept. of Artificial Intelligence, University of Edinburgh, 1996.

23. P. Maes, M. Matarić, J-A. Meyer, J. Pollack, H. Roitblat, and S. Wilson, editors. *From Animals to Animats: Proceedings of the Fourth International Conference on Simulation of Adaptive Behavior*, Cambridge, MA, 1996. MIT Press/Bradford Books.

24. M. Matarić and D. Cliff. Challenges in Evolving Controllers for Physical Robots. *Robotics and Autonomous Systems*, 19(1):67–83, 1996.

25. J. A. Meyer, H. L. Roitblat, and S. W. Wilson, editors. *From Animals to Animats: Proceedings of the Second International Conference on Simulation of Adaptive Behavior*, Cambridge, MA, 1993. MIT Press/Bradford Books.

26. O. Michel. *Khepera simulator Package*. LAMI, Swiss Federal Institute of Technology in Lausanne, Version 2.0 edition, 1996. Freeware mobile robot simulator downloadable from http://diwww.epfl.ch/lami/team/michel/khep-sim/.

27. F. Mondada and E. Franzi. *K213 Vision Turret User Manual*. K-Team S.A., 1028 Préverenges, Switzerland, Version 1.0 edition, 1995.

28. F. Mondada, E. Franzi, and P. Ienne. Mobile robot miniaturization: A tool for investigation in control algorithms. In T. Yoshikawa and F. Miyazaki, editors, *Proceedings of the Third International Symposium on Experimental Robotics*, pages 501–513, Tokyo, 1993. Springer Verlag.

29. F. Morán, A. Moreno, J. J. Merelo, and Chacón P., editors. *Proceedings of the Third European Conference on Artificial Life*, Berlin, 1995. Springer-Verlag.

30. J-D. Nicoud and O. Matthey. Developing intelligent micro-mechanisms. In *Proceedings of the Conference on Micromechatronics and Human Science*, Nagoya, Japan, 1997. IEEE Press.

31. S. Nolfi, D. Floreano, O. Miglino, and F. Mondada. How to evolve autonomous robots: Different approaches in evolutionary robotics. In R. Brooks and P. Maes, editors, *Proceedings of the Fourth Workshop on Artificial Life*, pages 190–197, Boston, MA, 1994. MIT Press.

32. D. Parisi, F. Cecconi, and S. Nolfi. Econets: Neural networks that learn in an environment. *Network*, 1:149–168, 1990.

Blurred Vision: Simulation-Reality Transfer of a Visually Guided Robot

Tom MC Smith

Centre for Computational Neuroscience and Robotics
School of Cognitive and Computing Sciences
University of Sussex
Brighton BN1 9SB, UK
toms@cogs.susx.ac.uk

Abstract. This paper investigates the evolution of robot controllers utilising only visual environment input data, capable of performing a *hard* task, playing football, in the real world. The techniques of *minimal simulation*, where the robot controller is forced to ignore certain features through making those features *unreliable*, are used to construct a noisy simulated environment for a robot with an onboard vision system. Robot control structures evolved in this simulation are then transferred on to a real robot, where the behaviours shown in simulation are displayed in the real world. In the experiment presented, finding a tennis ball and pushing it towards a goal, good controllers capable of performing the same behaviours in simulation and in the real world, are evolved only once sufficient unreliability is incorporated into the simulation. The success in evolving in simulation a robot controller incorporating only *distal* visual environment input data and displaying the same behaviours in both simulation and the real world, goes some way to addressing the argument that evolution is suitable only for toy problems.

"Some people believe football is a matter of life and death. I'm very disappointed in that attitude. I can assure you it is much, much more important than that."

Bill Shankly

1   Introduction

To operate well in the real world[1], embodied agents must utilise long distance sensor information. The incorporation of such *distal* information, as opposed to local proximity detector information, e.g. infra-red, enables the utilisation of far more sophisticated strategies [1]. This paper explores the evolution of sensorimotor controllers for robots operating autonomously in a real environment

---

[1] In the sense of not only the real *physical* world, but also an environment not artificially constructed for the robot.

with which they can *interact*, utilising sensor input provided by an onboard vision system; a difficult task previously avoided in the evolutionary robotics field.

The techniques of *minimal simulation* [6, 7] are used to construct a simulation for a robot sensorimotor control system incorporating *only* visual input data returned by an onboard camera. The experiment, a simplified version of football, requires the robot to find a bright ball in a dark arena and push it towards a striped goal. Behaviours evolved in simulation are successfully transferred to the real world, but only once sufficient unreliability is introduced to the simulation.

Section 2 discusses previous research in evolving visually guided robot controllers. Section 3 outlines the problems with evolution in both the real world and simulation, and introduces the techniques of minimal simulation. Section 4 outlines the robot and vision system used, while section 5 describes the minimal simulation of such a visually guided robot. Section 6 outlines the control structures and evolutionary process used, and section 7 describes the robot football experimental setup. Finally, section 8 summarises the experimental results, with discussion in section 9.

2 Evolved Visual Controllers

Previous research in evolving visually guided controllers has tended to concentrate on simulation, avoiding the difficulties inherent in using real world vision. Floreano and Mondada [4] investigate a predator-prey scenario using a simulation of the same Khepera robots and vision system used in this paper (see section 4 for a description). However, the vision system is only used to return the position of the darkest pixel in the field of view, and no evaluation of the controllers in reality is described (for an exploration of transferring controllers utilising this kind of visual input from simulation to reality, see [12]). Cliff *et al* [1] evolve controllers capable of finding the centre of a room in simulation, but again no evaluation of behaviour in the real world is described.

Other approaches include the Sussex gantry robot, "partway between a physical mobile robot ... and simulation" [5]. Jakobi [6, 7] has extended work by Harvey *et al* [5], successfully evolving controllers in a minimal simulation, capable of distinguishing shapes even in "disco" lighting conditions. However, neither the image processing and sensorimotor control are performed onboard the gantry robot.

As far as I am aware, no work on the evolution of visually guided real world robot controllers has been undertaken where the robots are capable of interacting with their environment, beyond being able to move around in that environment. This paper describes the successful construction and analysis of such a setup, where the robot must rely purely on visual sensor information provided by an onboard camera, to find and push a ball.

3   Minimal Simulation

The issue of robot controller evolution in the real world versus evolution in simulation has traditionally been the trade-off between realism and time[2].

Evolution in the real world suffers from the problem of the inordinate evaluation time required. With a large population evaluated over number of generations, the real time evolution cost can be prohibitively large. Floreano and Mondada [2] required 65 hours to evolve efficient collision-avoidance controllers, and ten *days* to evolve learning controllers to perform the same task [3].

By contrast, it is argued that simulation cannot realistically model the features required for robust operation in the real world; robots evolved in simulation may completely or partially fail in the real world - the so-called "reality gap" [8]. Attempts to increase the simulation complexity merely result in expenditure of vast amounts of modelling and computing time [9].

Jakobi's *Radical Envelope of Noise* hypothesis [6, 7] outlines a new approach to the problems of evolution in simulation. The contrast is drawn between the *base set aspects* of the situation, those that may have some bearing on the robot behaviour, and the *implementation aspects*, those which must not be allowed to affect behaviour.

Base set aspects are those simulation features present in the real world upon which the robot behaviour might be based. For robustness, these are modelled noisily. Implementation aspects are those simulation features either not present in the real world (perhaps arbitrary regularities), not thought relevant, or not easily modelled. Instead of arbitrarily setting these aspects, or allowing some random distribution, these must be made *unreliable*, e.g. randomly set for each trial to one of 'on', 'off', 'big', 'little', 'randomly distributed', etc. The only practical evolutionary strategy is to ignore them completely.

The key point is that the base set need not be comprehensive, or particularly accurately modelled; a *minimal simulation* base set will only evolve controllers dependent on a small number of features, possibly not able to exploit the full real world situation, but able to cross the reality gap. Section 4 introduces the robot and vision system.

4   The Robot and Vision System

The work described in this paper used the standard research minirobot Khepera [10], equipped with the plug-in K213 vision module. The vision module returns a one-dimensional line-scan of light intensity values across a 36° arc. The full resolution returns 64 values across this arc; this was reduced by a factor of four for the controller input, scaled to range from 0 to 1 before processing was applied.

---

[2] The difficulties described here refer only to the phenotype evaluation part of the evolutionary process. The genotype manipulation (breeding, mutation and so on) will clearly be carried out by computer.

The vision system incorporates a hardwired 'iris', whereby overall incident light intensity is used to alter the time over which individual pixel input intensities are integrated. Thus the visual input for objects filling the field of view cannot be used to decide how bright that object is - a white background looks much like a black background[3].

The real world is both complex and noisy, and visual input data invariably reflects this; the data returned from the Khepera vision turret is no exception. A variety of image processing techniques exist both to emphasise the main image features present and to reduce noise. However, the constraint that any image processing must be performed onboard the robot in real time removes the possibility of using such computationally expensive techniques as Laplacian transformation of Gaussian convoluted data [11]. Any processing must be 'quick and dirty'.

The processing algorithm used here is a weighted sharpening operator, where the processed value $g(x)$ is the difference between the unprocessed value $f(x)$, and a simple weighted sum of the surrounding three values, see equation 1:

$$g(x) = f(x) - \sum_{i=-1}^{i=1} A(i)f(x+i) \tag{1}$$

$$A(-1) = A(1) = 0.25$$
$$A(0) = 0.5$$

5 Minimal Simulation of a Visually Guided Robot

This section puts into practice the theory discussed in section 3; how to identify the base set and implementation set aspects necessary for the construction of a visually guided robot minimal simulation.

5.1 Simulating the Vision System

Only one feature was identified as a minimal simulation base set aspect in the vision system; how high and low light intensity levels affect the pixel inputs *when viewed at the same time*[4].

From characterisation of the K213 vision system (described more fully in [12]), two implementation aspects were identified:

- The performance of individual pixels given identical inputs is not reliably the same. This was varied across trials by adding different values to each of the pixel inputs, chosen from either a. nothing, b. Gaussian distributed, or c. uniform distributed random values.

---

[3] When pushing a tennis ball, the ball fills the field of view, so the robot does not 'see' the ball at all.

[4] In general it is not possible to decide whether a given *uniform* background is light or dark from the visual data alone, due to the iris affecting input integration time.

– The angle of view of individual pixels[5] was also varied slightly from an ideal setting for each of the pixels, by adding either a. nothing, b. Gaussian distributed, or c. uniform distributed random values.

5.2 Simulating the Robot

Again, only one base set aspect was identified; how the motors affect the robot location. The new robot location is calculated on the basis of the left and right motor speeds over the last two time steps, and the time taken for one update of the neural network[6]. The wheel speed is chosen randomly between the wheel speeds on the last two time steps, thus producing a random momentum element. This momentum was included once it was seen that early controllers were failing in the real world, spinning past the tennis ball before acting - in the simulation, wheel speeds are set instantaneously, which is clearly not the case in the real world.

The simulation is crude in that a lookup table is used to avoid computationally expensive angle calculations when relating the robot orientation to direction of motion. Finally, noise is added to the robot position and orientation.

The robot interaction with a wall is extremely hard to model accurately, yet the simulation needs to be set up in such a way as to discourage the robot to coming close to the walls. The solution is to make the robot-wall interaction an unreliable implementation aspect - controllers are allowed to hit the wall without being penalised, but they will not be able to rely on the effects of such interaction. The unreliability implemented here provided four possible consequences of hitting the wall:

– Robot does nothing, stopping still.
– Robot is bounced back a random distance, and spun in a uniform random orientation.
– Robot is moved in a uniform random direction, and spun in a uniform random orientation.
– Robot is moved in a Gaussian distributed random direction, and spun in a Gaussian distributed random orientation.

As the robot has no way of knowing how close it is to the wall, the effect is to discourage motion unless viewing something 'bright' which is definitely not the wall; good controllers cannot hit the wall and manage to track the ball or push it towards the goal consistently.

---

[5] To minimise simulation time, inputs are not integrated over some area, but taken from a single ray along the centre of the pixel input angle. However, this angle is varied on each evaluation to prevent controllers using strategies based on fine differences in angle of view between the pixels.

[6] To cross the gap between simulation and reality, the time taken in simulation for one network update must be similar to the time taken in reality. Strategies evolved on the basis of moving a certain distance between updates may fail completely if they do not move the expected distance in reality.

5.3 Simulating the Physical Environment

The interaction of the robot with the ball is clearly crucial to the experiment. However, the ball used, a standard tennis ball, is clearly far from ideal with respect to elastic collisions, smooth rolling, and other easily computed factors. Specified as a base aspect, the ball's motion after collision is in the direction along the robot-ball centres, and is moved by a factor dependent on the robot's velocity. Uniform noise is added to both the direction of motion and distance moved (the ball does not roll, but 'jumps' - the distance moved is moved in one simulation time step).

Finally, the overall environment was varied across trials; the size and colour of the objects, the background lighting values, initial robot position, and the number of stripes on the goal. However, the actual number and shape of the objects - two rectangular goals, four walls forming a rectangular arena, and one round ball - were kept constant.

6 Genetic Algorithms and Neural Networks

6.1 The Genetic Algorithm Scheme

The work described in this paper used a *distributed genetic algorithm*, in which each solution is considered to occupy a unique position on a two-dimensional toroidal grid. Initially, the grid is seeded with random genomes, each mapping on to a single neural network controller (section 6.2 gives details of the network, and the genotype to phenotype mapping). Each solution is evaluated in terms of its 'footballing' ability (see section 7) and the main program loop entered.

On each generation, the algorithm iterates *PopulationSize* times, choosing each time a random location on the solution grid. A *mating pool*, consisting of the current location plus the neighbouring eight, is set up centred on the randomly chosen grid location, and the mating pool solutions ranked in order of fitness. Rank-based roulette selection is used to find two parents, and breeding produces an offspring solution which is evaluated and placed back in the mating pool at a point chosen by inverse rank-based roulette selection. Breeding is either asexual, in which the child solution is a mutated version of the first parent, or sexual, where one point crossover of the two parent solutions produces a child solution which is then mutated.

6.2 The Neural Network Sensorimotor Control Structure

The solution genotype is a direct encoding of the robot sensorimotor control network. The neural network used was a fixed architecture three-layer neural network, see figure 1. The 16 visual inputs, compass[7], and two recurrent motor

---

[7] The compass input to the network is a binary coding indicating whether the robot is facing its own goal or the opposing goal. This binary is based on an orientation of the robot calculated on the motor outputs of the neural network (thus available to the robot). However, this soon becomes unreliable due to robot-ball and robot-wall collisions, and noise on the robot position update.

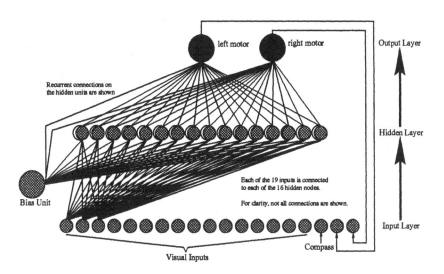

Fig. 1. Feed-forward neural network controller

connections, are fully connected to each of the 16 hidden layer neurons. The hidden inputs also receive input from a bias unit, and recurrent connections from their own output on the previous time step. Finally, each of the 16 hidden units is connected to the left and right motor outputs.

The transfer function relating a neuron input to its output is a simple thresholded linear function with random noise added to the input ($\pm$ 0.05). Initially, each bit on the genotype is randomly seeded in the range -0.5 to 0.5, but not constrained afterwards.

7 Robot Football

7.1 Experimental Setup

The football setup consists of an arena, 105cm long by 68cm wide[8]. The walls and floor are black, while the 30cmx15cm goals are black and white striped (stripe width 3cm). The football is a yellow tennis ball, which the robot must find and push towards the opposing goal. Evaluation of each robot network controller was simulated for 100 seconds in the arena, stopping if a goal was scored.

Initially, the ball is placed at the centre of the arena, with the robot placed randomly and facing the opposing goal. On each time step, the visual input is found (including the environment and vision system noise and unreliability) and the network output used to update the robot position.

7.2 Fitness Evaluation

Robot controllers were evaluated for 'footballing' ability in the minimal simulation described in section 5. Equation 2 shows the fitness evaluation function. For

---

[8] Values shown are mean values - the actual values vary on each evaluation.

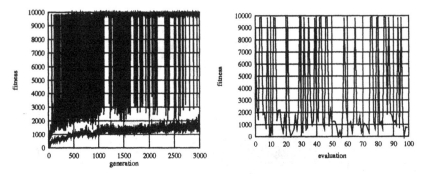

Fig. 2. Fitness evaluation of controllers, a. best and average fitnesses over generations, and b. fitness of best 3000th generation controller over 100 evaluations.

each controller, ten simulated evaluations were carried out, and fitness returned as the median value.

$$fitness = \frac{TimeLookingAtBall}{TotalTime} + f(ClosestDistToBall) +$$
$$\sum_{N} g(HitBall) + h(BallToGoalDistance) \quad (2)$$

$f(x) \propto x$ Score for moving towards ball
$g(x)$ Score for hitting ball
$h(x) \propto (C - x)^2$ Score for moving ball towards goal
$N \leq 20$ no. of ball-robot collisions rewarded
If goal scored, $fitness = GoalScore - TimeTaken$

The fitness evaluation used here explicitly rewards a series of behaviours identified as being necessary to scoring goals - see the ball, go towards the ball, and push the ball towards the goal. Evaluation of how many goals were scored or how close the ball is to the goal, which implicitly reward sub-behaviours such as seeing and finding the ball, were found not to produce good solutions.

8 Results

Controllers displaying efficient ball finding skills *in the real world* were consistently evolved in simulation over 3000 generations, taking roughly thirty hours on a SPARC Ultra 2 (0.036% of the 9.5 years of real world evaluation time that would be required). Figure 2 shows the evaluated fitness over generations, and evaluated fitness of the best individual over several evaluations, while figure 3 shows four different simulated evaluations.

Controllers start scoring around generation 50, but even by generation 3000, the best controller only scores in roughly 25% of evaluations. Note, scoring a goal obtains a fitness of $10000 - TimeTaken$, while pushing the ball, but not towards the goal, scores roughly 300.

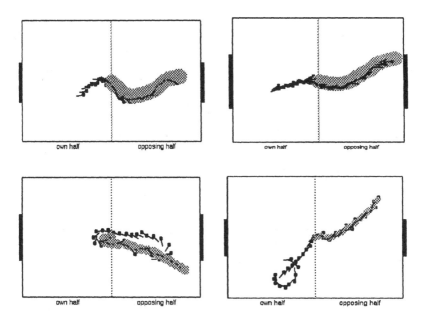

Fig. 3. Four simulation evaluations. Note, Khepera robot and orientation is indicated by pin, and tennis ball is grey circle. Evaluation is shown every 2 updates (roughly 0.5 seconds). Arena size may change on each evaluation.

20 evaluations were carried out in the real world, and in all but two the robot span on the spot until the ball was 'seen', before heading straight towards the ball, adjusting to keep it in the centre of the field of view. The robot continued to push the tennis ball, turning to follow if the ball rolled to one side. The two failures occurred when the robot initial position was near its own goal; in both cases the robot headed straight for the goal and became stuck against the wall.

The video footage of the robot (figures 4 and 5) shows the evolved controllers exhibit extremely good ball finding skills in the real world. It should be emphasised that this is not a trivial task, especially given the constraints of all processing and image capture taking place onboard the robot.

The vision system failed to 'see' the ball at all if lighting levels were too low[9]; in practice it was found necessary to use overhead angle-poise lights. However, the use of different coloured balls was coped with by the controller, so long as there was sufficient contrast between the ball and background.

The robot scored four goals from the 18 evaluations where the ball was found - in two cases the robot started with the ball nearly lined up between it and the goal, but in the other two cases the robot approached the ball from one side, managing to line it up with the goal before scoring. However, it is unclear to what extent the controller is actively 'scoring goals', and not just pushing the ball

---

[9] The vision turret simply returned uniform values when this was the case.

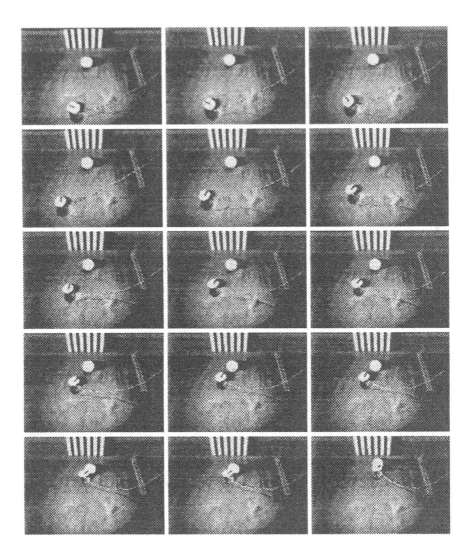

Fig. 4. Khepera robot turning to find the ball, and moving towards it, despite the goal also being placed nearby. Once the robot reaches the ball, it carries on pushing the ball forward. NB: the Khepera's forward orientation is shown by the stripe on top of the robot; the 30cm ruler at the top right shows scale; the serial cable provides only power - the robot is not connected to a computer.

in that direction - the unreliability of a compass based on wheel motion makes scoring an extremely difficult task. Clearly, more effective navigation requires a better compass; digital compasses providing accurate orientation are widely available and could be wired in as an external input to the neural network.

Qualitative evaluation of network activity shows certain key neurons acting as bright object detectors - these enable the robot to centre the ball extremely quickly and move forwards. However, it is unclear how the network filters out

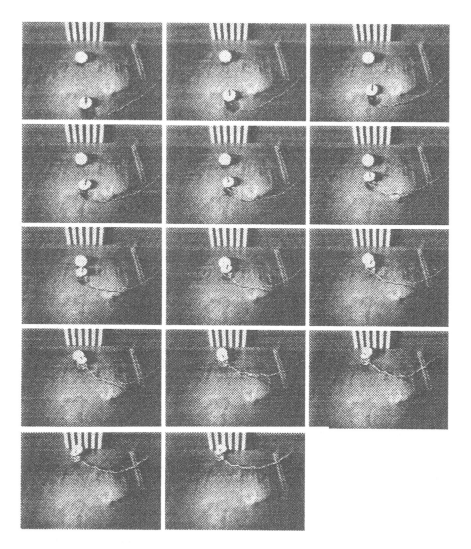

Fig. 5. Khepera robot scoring 'penalty' - note the slight turns made in frames 4 and 5 to line the ball up dead-centre. NB: the Khepera's forward orientation is shown by the stripe on top of the robot; the 30cm ruler at the top right shows scale; the serial cable provides only power - the robot is not connected to a computer.

smaller bright objects, such as the goal stripes, or how the controller recognises that it is pushing the ball once the ball fills its field of view.

9 Discussion

The potential power of embodied agents lies in their ability to assimilate and interact with the real world. This paper has shown that evolution is capable of producing such robot controllers, able to find and move an object guided only by onboard visual sensors.

The minimal simulation constructed of the robot, vision system and visual environment produced an evaluation of robot footballing ability that was accurate enough to successfully transfer controllers to the real world. Controllers were allowed to base behaviour only on certain aspects of the simulation; other features were ignored. Despite the basic nature of the onboard vision system, a one-dimensional linescan with hardwired contrast gain, and the constraint of all processing being onboard in real time, controllers evolved to produce extremely good ball finding and pushing behaviours.

Further research is currently investigating allowing the neural network architecture to be genetically determined, the use of more sophisticated vision systems and image processing, and the use of less explicit fitness evaluation functions. However, the results presented here demonstrate that artificial evolution is capable of scaling up to real world interactive tasks requiring visual environment data; vital if evolutionary robotics is to produce controllers going beyond toy problems.

10 Acknowledgements

Many thanks to Phil Husbands for insightful supervision, and to Nick Jakobi for introducing me to robots. Finally, thanks to Anil Seth, Andy Philippides and Matthew Quinn for constructive comments and proof-reading.

References

1. D.T. Cliff, P. Husbands, and I. Harvey. Evolving visually guided robots. In J.-A. Meyer, H. Roitblat, and S. Wilson, editors, *Proceedings of the Second International Conference on Simulation of Adaptive Behaviour*. MIT Press/Bradford Books, 1993.
2. D. Floreano and F. Mondada. Automatic creation of an autonomous agent: Genetic evolution of a neural-network driven robot. In D. Cliff, P. Husbands, J. A. Meyer, and S. W. Wilson, editors, *Animals to Animats 3: Proc. 3rd International Conference on Simulation of Adaptive Behaviour*. MIT Press/Bradford Books, 1994.
3. D. Floreano and F. Mondada. Evolution of plastic neurocontrollers for situated agents. In P. Maes, M. J. Mataric, J. A. Meyer, J. Pollack, and S. W. Wilson, editors, *From Animals to Animats 4; Proc. 4th International Conference on Simulation of Adaptive Behaviour*. MIT Press/Bradford Books, 1996.
4. D. Floreano and S. Nolfi. Adaptive behaviour in competing co-evolving species. In P. Husbands and I. Harvey, editors, *Fourth European Conference on Artificial Life*. MIT Press/Bradford Books, 1997.
5. I. Harvey, P. Husbands, D. T. Cliff, A. Thompson, and N. Jakobi. Evolutionary robotics: The Sussex approach. *Robotics and Autonomous Systems*, 20:205–224, 1997.
6. N. Jakobi. Evolutionary robotics and the radical envelope of noise hypothesis. *Journal of Adaptive Behaviour*, 6, 1997.
7. N. Jakobi. Half-baked, ad-hoc and noisy: Minimal simulations for evolutionary robotics. In P. Husbands and I. Harvey, editors, *Fourth European Conference on Artificial Life*. MIT Press/Bradford Books, 1997.

8. N. Jakobi, P. Husbands, and I. Harvey. Noise and the reality gap: The use of simulation in evolutionary robotics. In F. Moran, A. Moreno, J. Merelo, and P. Chacon, editors, *Advances in Artificial Life: Proc. 3rd European Conference on Artificial Life*. Springer-Verlag, 1995.

9. M. J. Mataric and D. T. Cliff. Challenges in evolving controllers for physical robots. *Robotics and Autonomous Systems*, 19(1):67–83, 1995.

10. F. Mondada, E. Franzi, and P. Ienne. Mobile robot miniaturization: A tool for investigation in control algorithms. In *ISER '93*, Kyoto, Japan, October 1993.

11. R. J. Schalkoff. *Digital Image Processing and Computer Vision*. John Wiley and Sons, 1989.

12. T.M.C. Smith. Adding vision to Khepera: An autonomous robot footballer. Master's thesis, School of Cognitive and Computing Sciences, University of Sussex, 1997.

Learning to Move a Robot with Random Morphology

Peter Dittrich[1] , Andreas Bürgel[1] and Wolfgang Banzhaf[12]

[1] Dept. of Computer Science, University of Dortmund,44221 Dortmund, Germany
http://ls11-www.informatik.uni-dortmund.de
dittrich | buergel | banzhaf@LS11.informatik.uni-dortmund.de
[2] Presently at: International Computer Science Institute, Berkeley, CA, 94708

Abstract. Complex robots inspired by biological systems usually consist of many dependent actuators and are difficult to control. If no model is available automatic learning and adaptation methods have to be applied. The aim of this contribution is twofold: (1) To present an easy to maintain and cheap test platform, which fulfils the requirements of a complex control problem. (2) To discuss the application of Genetic Programming for evolution of control programs in real time. An extensive number of experiments with two real robots has been carried out.
Keywords genetic programming, real-time robotics, random morphology robot, hardware evolution

1 Complex Bio-inspired Robots

Conventional industrial robots are designed in such a way that a model can be derived easily and the inverse kinematic can be calculated. In operation, the inverse kinematics is used to compute the trajectory for movement between given points in the working area of the robot. Connections between actuators are made as sticky as possible to yield (near) linear behaviour [7]. Perception relies on sense-model-plan-act cycle, where for planning a mostly predefined model of the system is required.

For the development of robots which are inspired by biological systems [1] "controllability" is not a primary design principal. Thus, their actuators are mostly dependent. A model usually does not exist, is very hard to derive or too complex so that a model-based calculation of motor commands requires too much time for reactive tasks. There is no obvious optimal control strategy for a desired action (e.g. movement) because of the complex interdependencies of all the actuators and a non-linear feedback from the environment. Examples of (at least partially) bio-inspired robots are modular robots like the robot snake [2, 3] and robot fishes like the robot tuna build at MIT Ocean Engineering [4].

But even if a model exists, a robot can get into a situation where this model is not valid anymore, e.g. through malfunctioning of parts. If the control strategy is based on the model and an unexpected error occurs (e.g. the breaking of a

---

[1] Shortly called: bio-inspired robot.

joint between two actuators) the model breaks down and the control strategy is likely to fail [2].

In this case a learning mechanism would be very useful that is able to generate a new control program adapted to the new situation.

In other words, every robot can turn into what we call a random morphology ("RM-") robot, where "RM-robot" refers to a robot with an arbitrary, complex architecture. In this paper, we study a 6-servo robot (see Figure 1) as an RM-robot. The term RM-robot does, however, *not* imply non-deterministic behaviour.

If the robot is on its own, an adaptation mechanism is needed which is able to cope with an unexpected architecture and which makes as few assumptions as possible about the hardware. In the following we will (1) present an easy to maintain real robot platform to test such mechanisms and (2) discuss Genetic Programming (GP) as a mechanism to cope with a RM robot.

It should be noted that the RM robot is also inspired by Sims's work on evolving morphologies [1]. From this point of view the RM robot can be seen as a step towards a physical instantiation of Sims's virtual box creatures.

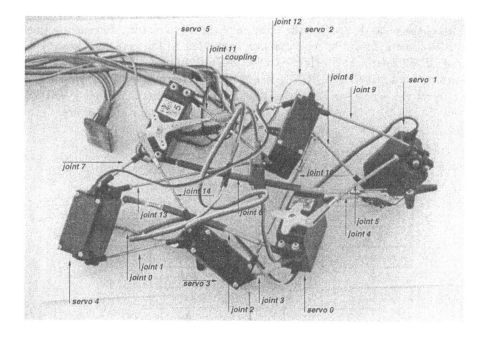

Fig. 1. The mechanics of the random morphology (RM-) robot. It consists of six conventional servos coupled arbitrarily by thin metal joints.

---

[2] Strictly speaking. the breaking of a joint is not an unexpected error, because we have already expected it. Thus, its very hard, maybe impossible, to model unexpected errors.

2 The RM-Robot

Actuators:

The RM robot is composed of a couple of servos which are connected arbitrarily (randomly). The servos are conventional cheap RC (remote control) servo motors available for hobby air planes and cars. These devices possess a complete servo system including: motor, gear box, feedback device, servo control circuitry, and drive circuit. The connections are made by brass poles also available for hobby modelling. They can be easily connected to the servos, thus one can set up or change an architecture quickly, which should be useful for evolutionary experiments in hardware. The complexity of the mechanics can be increased by connecting poles and servos with springs.

Sensors:

Movement of the RM-robot is measured by a computer mouse device, mechanically connected to the robot. This device allows precise measurement of motion in the 2-D plane. There are also light detecting sensors which are, however, not considered in this contribution.

Control:

The servos are controlled by a pulse signal that occurs at about 50 Hz. The width of the pulse determines the position of the servo motors. To generate this signal we use a simple micro-controller, connected to the host computer by a serial RS232 interface. The host is a PC running LINUX which is fast enough even without a real-time LINUX kernel. A piece of interface software was written to control the servos via the serial RS232 line and to measure analog voltage input via a A/D PC card. Figure 2 shows the overall system architecture.

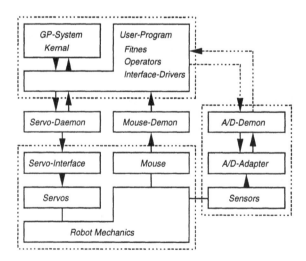

Fig. 2. Overview of the system architecture.

Discussion:

The system is composed of conventional and cheap parts. It is easy to maintain and to construct. It provides an interesting and very flexible test platform for adaptive and learning algorithms that have to cope with complex, unexpected architectures. Its main limitation is the external control by a desktop computer, which is not a problem from a scientific point of view. A wireless, fully autonomous version would make experiments and demonstrations simpler. At the moment the robot has to be watched constantly by an experimentator because it may interact with its wire which would bias the results.

3 Evolving Control Programs

In this section we will show how control programs can be generated by an evolutionary process using Genetic Programming [10,5]. There are various ways how Evolutionary Algorithms (EA) can be applied to generate or optimize robot controllers [17,6,16]:

1. The fitness evaluation can be performed by a simulator, as in [11]. The advantages of this method are: Different controllers can be tested under exactly the same environmental conditions. A simulation is usually cheaper and faster than a real robot. Stability, robustness and correction correctness of solutions be proven – although not in general – which is important for an industrial application. On the other hand, problems with this approach are: A model must be available and this model or simulation might have artifacts, e.g. deadlocks. The individuals might exploit simplification or artifacts in the model [1].[3] As a result, the evolved controller might not work (reliably) on the real robot.
2. The fitness can be evaluated using the real robot [15,12,13]. The major qualitative differences to the simulator-based evaluation are: Fitness evaluation is now a stochastic process and real-world time plays an important role. In addition to the robot learning system, a changing environment must be taken into account where time flow is not synchronised with the learning system. For instance, changes of the environment might be slowed-down or stopped in a simulation whereas this is not possible in the real world. An advantage is that one may encounter "unexpected errors" which do not appear in simulations.
3. A combination of (1) and (2) can be used [17]. A controller generated by a simulation is fine tuned on the real robot. To evolve complex control systems the task is usually divided into subtasks (e.g. behaviours) which are independently evolved either by simulation or using the real robot. The behaviours are combined by an action selection mechanism which can be evolved, too [9].

Fitness evaluation can also be characterised by a time scale:

---

[3] This effect can be put to use when testing models or simulators.

1. Global or goal-oriented fitness evaluation: The robot is run for a long time during which it is able to reach the desired goal once, or even many times starting from the same or different positions. Fitness can be easily derived from a measure how well the goal has been reached. This method is usually used in simulations.

2. Local fitness evaluation: The robot is run only for a very short time (in [12] only a fraction of a second). The fitness evaluation will be much faster, but defining the fitness function on local actions such that the global goal will be reached is more difficult than in the previous case.

Of course, it cannot be stated in general that one method is better than the other. It is very probable that in most applications a combination of different methods is favourable. Here, we shall concentrate on one method in Genetic Programming which we will use with only a few small modifications.

The motivation for using GP is: (1) We would like to examine whether or not GP is able to learn complex robot movements in real-time and uncover the benefits and limitations of this approach. (2) GP produces automatically programs. A program is a very flexible and most commonly used representation of a computable function.

3.1 The Genetic Programming System

The learning method is a conventional steady-state tree-based GP algorithm using a local fitness evaluation on a real robot with the following settings.

We use local fitness evaluation with a steady-state algorithm. An individual will control the robot only for a short while. For this it is executed n_e times. Its performance is measured as the robots advance in the desired direction. Backward movement will be punished by a factor of two and stagnation will be equal to the worst backward movement so far. The Fitness is evaluated at creation time not during the selection process. In tournaments, an individual is selected by drawing an number of individuals uniformly distributed from the population and choosing the individual with the best (or worst) fitness value. As in in conventional GP one subtree in each parent is selected randomly and exchanged. In mutation, each node is mutated by a probability $p_{nodeMutation}$. For this the node is replaced by a node randomly selected from the set of nodes with the same arity. So, a terminal is always replaced by another terminal. The arity of a node will never be changed by mutation.

The set of functions in Tab. 1 is explained in more detail in Tab. 3. The terminal set is explained in Tab. 3.

3.2 Algorithm

Because in real-time evolutionary learning it is important exactly when and how the fitness is evaluated, the algorithm is given in more detail below. Implementation is based on *gpquick* [14].

Objective	Find a program that moves the robot straight on as far as possible
Raw fitness	The sum of pixels the mouse pointer travels in a desired direction minus the sum of pixels the mouse pointer travels in the opposite direction. (See text.)
Fitness	Equal to raw fitness, except in the case when the raw fitness is zero the fitness is equal to the worst fitness so far encountered.
Executions per fitness evaluation	$n_{repetitions} = 4$
Terminal set	GETSERVO0, GETSERVO1, GETSERVO2, GETSERVO3, GETSERVO4, GETSERVO5, GETSERVO6, CONST
Function set	ADD, SUB, MUL, DIV, SINE, DELAY, SETSERVO0, SETSERVO1, SETSERVO2, SETSERVO3, SETSERVO4, SETSERVO5, SETSERVO6, IF, IFLTE, SEQUENCE2, PROG4
Population size	$M = 50, 100$
Maximal number of nodes	$l_{max} = 100, 200$ nodes
Probability of mutation	$p_m = 0.13$
Probability of node mutation	$p_{nodeMutation} = 0.99, 0.15$
Probability of crossover	$p_c = 0.86$
Probability of reproduction	$p_r = 0.01$
Tournament size for genetic operators	$T_r = 4$
Tournament size for replacement	$T_k = 2$
Termination criteria	running time excess or decision by experimenter

Table 1. Koza tableau of the evolution of motion control programs for the RM-robot.

ADD(a,b), SUB(a,b), MUL(a,b), SINE(a)	normal arithmetic operation
DIV(a,b)	protected division, returns 1 if $b = 0$
DELAY(t)	delays the execution of the program for t time steps
SETSERVO0(a), SETSERVO1(a), ..., SETSERVO5(a)	Commands servo to position a. No delay is executed. Returns current position of the servo. Values $a > 127$ and $a < -127$ result in a maximal left or right turn, respectively.
IF(a,b,c)	if $a > 0$, returns b, else c
IFLTE(a,b,c,d)	if $a \leq b$, returns c, else d.
SEQUENCE2(a,b)	evaluates a, then b, returns result of b
PROG4(a,b,c,d)	evaluates a, then b, then c, then d, returns result of d

Table 2. Function set of the GP system.

GETSERVO0, GETSERVO1, ..., GETSERVO5	returns current position of servo n
CONST	a fixed random constant out of [-127, 127]

Table 3. Terminal set of the GP system.

Initialisation:

1. Generate a random population P of size M.
2. For each individual in P, evaluate its fitness.

The GP execution cycle:

1. Delete one individual from P, which is selected by a tournament of size T_k (worst individual from T_k randomly drawn individuals).
2. Draw a random number x out of $[0, 1]$.
3. If $x < p_c$:
   Select I_1 from P by a tournament of size T_r.
   Select I_2 from P by a tournament of size T_r.
   Create offspring I_o via crossover of I_1, I_2.
4. If $p_c < x < p_c + p_m$:
   Select I from P by a tournament of size T_r.
   Create offspring I_o via mutation of I.
5. If $p_c + p_m < x < p_c + p_m + p_r$:
   Select I from P by a tournament of size T_r.
   Create offspring I_o with $I_o = I$.
6. Evaluate fitness for offspring I_o.
7. Add offspring I_o to population P.

A tournament of size T_r means that T_r individuals are randomly drawn from the population and the best according to its fitness is selected. Note that the fitness of each individual has been evaluated before, either during initialisation or in step (6). Although this is an efficient procedure, it might become a significant problem, because fitness values can become obsolete. The above algorithm will fail if the environment changes very quickly, which is not the case, however, in the experiments described here. The problem could be overcome by evaluating the fitness again for each individual during a tournament, like in [12].

Evaluation of fitness f of individual I:

1. Clear mouse event queue.
2. Execute $n_{repetitions}$ times Individual I. This creates a mouse event queue of size n_q.
3. For each mouse event i set E_i to the distance the pointer moved in the desired direction represented by this event. Backward movement results in a negative E_i.
4. $f \leftarrow \sum_{i=1}^{n_q} E_i$ (Variant B: $f \leftarrow (\sum_{i=1,E_i>0}^{n_q} E_i)^2 - 2(\sum_{i=1,E_i<0}^{n_q} E_i)^2$)
5. $f_{min} \leftarrow \min(f_{min}, f)$.
6. if $f = 0$, set $f \leftarrow f_{min}$.

3.3 Results

Before describing the results with the 6-servo RM robot, preliminary experiments with a 3-servo robot will be shortly summarised.

3.4 Preliminary Experiments

A series of 55 experiments has been performed with the preliminary 3-servo robot shown in Figure 3. An experiment is a run of the robot with the algorithm described above. The preliminary robot consists of two servos able to retard the front wheels and the rear wheels, respectively. The third servo can bend the robot in the middle. In these experiments, we were able to evolve programs for movement in the plane and even on a gradient. The gradient is much more difficult because the two servos functioning as breaks must be synchronised very accurately. Furthermore, the experiments showed that

- punishment of zero-movement decreases the average number of non-moving individuals and increases convergence speed.
- learning how to delay is important. However, increasing the DELAY concentration only (no. of DELAY operations in the population) is not enough. The arguments of DELAY have to be large. This is achieved by multiplication of large numbers or (more seldom) by division by very small numbers.
- a MUL operation as an argument of a SETSERVOn operation usually creates an overflow (a value greater 127) that results in a maximum turn of the servo.

It could not been convincingly shown, that the overflow behaviour is exploited for protection against crossover and mutation.

3.5 Experiments with the 6-Servo Random Morphology Robot

A series of 32 experiments has been performed with the 6-servo RM-robot. As in the previous section an experiment is a run of the GP algorithm described above. A single run lasts several hours. Details can be found in Tab. 4 which gives an overview of all 32 experiments. There are 16 experiments with a population size $M = 50$ and max. individual size $I_{max} = 50$ and 16 experiments with $M = 100$ and $I_{max} = 100(500)$. In addition the operator set has been varied in order to explore the parameter space and to test the robustness of the GP system according to its setting. In each experiment the operator set consists at least of a core set $F_2 = \{SETSERVO0, SETSERVO1, \ldots, SETSERVO5, DELAY\}$ and an arithmetic function.

From these experiments, it becomes clear that the same system which learns to move the 3-servo preliminary robot is also able to move the 6-servo RM-robot. Although it is obvious that no significant conclusion can be made about which parameter setting is best, the following tendencies should be noted:

- Fitness evaluation using variant A (sum over E_i) together with a population size of $M = 100$ is better than using variant B and $M = 50$.

#	1997	f	M	l_{max}	Operator set	Tests	Bst	Wst	Avg
56	11.8	B	50	50	$F_2 \cup$ {ADD, SUB, MUL, DIV}	8000	69	-25	40
57	14.8	B	50	50	$F_2 \cup$ {ADD}	4350	20	-17	9
58	16.8	B	50	50	$F_2 \cup$ {ADD, SUB, MUL, DIV}	5150	52	-23	29
59	19.8	B	50	50	$F_2 \cup$ {ADD, SUB, MUL, DIV}	5000	54	-42	22
60	21.8	B	50	50	$F_2 \cup$ {ADD, SUB, MUL, DIV}	8000	45	-25	14
61	22.8	B	50	50	$F_2 \cup$ {ADD, SUB, MUL, DIV}	7000	35	-34	14
62	23.8	B	50	50	$F_2 \cup$ {ADD, SUB, MUL, DIV}	4400	40	-20	7
63	24.8	B	50	50	$F_2 \cup$ {ADD, SUB, MUL, DIV, IF, IFLTE, GETSERVOX}	7000	53	-29	23
64	25.8	B	50	50	$F_2 \cup$ {SUB, MUL}	5600	23	-25	9
65	25.8	B	50	50	$F_2 \cup$ {SUB, MUL}	1500	16	-22	3
66	25.8	B	50	50	$F_2 \cup$ {ADD, SUB, MUL, DIV, IF, IFLTE}	7000	32	-29	9
67	26.8	B	50	50	$F_2 \cup$ {ADD, SUB, MUL, DIV, IF, IFLTE}	2600	26	-29	10
68	27.8	B	50	50	$F_2 \cup$ {ADD, SUB, MUL, DIV, IF, IFLTE}	9100	33	-29	16
71	27.8	B	50	50	$F_2 \cup$ {ADD, MUL}	2000	22	-31	4
72	28.8	A	50	50	$F_2 \cup$ {ADD, MUL}	7850	22	-22	10
73	1.9	A	50	100	$F_2 \cup$ {ADD, SUB, MUL, DIV}	2850	20	-25	5
74	3.9	A	100	100	$F_2 \cup$ {ADD, SUB, MUL, DIV}	3800	136	-64	70
75	4.9	A	100	100	$F_2 \cup$ {ADD, SUB, MUL, DIV}	4000	22	-18	8
76	5.9	A	100	100	$F_2 \cup$ {ADD, MUL}	2500	180	-60	148
77	8.9	A	100	100	$F_2 \cup$ {ADD, MUL}	2000	115	-31	82
78	9.9	A	100	100	$F_2 \cup$ {ADD, MUL}	4000	20	-14	4
79	10.9	A	100	100	$F_2 \cup$ {ADD, MUL}	1800	135	-44	120
80	11.9	A	100	100	$F_2 \cup$ {ADD, SUB, MUL, DIV, IF, IFLTE}	3100	26	-23	14
81	13.9	A	100	100	$F_2 \cup$ {ADD, SUB, MUL, DIV, IF, IFLTE}	1700	19	-17	9
82	14.9	A	100	500	$F_2 \cup$ {ADD, SUB, MUL, DIV, IF, IFLTE}	4100	22	-21	13
83	15.9	A	100	500	$F_2 \cup$ {ADD, SUB, MUL, DIV}	3800	148	-24	127
84	17.9	A	100	500	$F_2 \cup$ {ADD, MUL}	1400	147	-33	108
85	18.9	A	100	100	$F_2 \cup$ {ADD, SUB, MUL, DIV}	1600	177	-14	168
86	19.9	A	100	100	$F_2 \cup$ {ADD, SUB, MUL, DIV, IFLTE}	3000	31	-25	13
87	19.9	A	100	100	$F_2 \cup$ {ADD, MUL, IFLTE}	2500	77	-22	68
88	22.9	A	100	100	$F_2 \cup$ {MUL, IFLTE}	3000	36		26

Table 4. Overview of the experiments with the 6-servo RM-robot. The table shows from left to right: experiment number, day of experiments realization, variant of fitness calculation, population size M, max. number of nodes per individual l_{max}, operator set, total number of fitness evaluation, best/worst/average fitness in final population. $F_2 = \{SETSERVO0, SETSERVO1 \ldots, SETSERVO5, DELAY\}$

Fig. 3. The preliminary robot with 3 servos. Two servos are used as brakes for the front and rear wheels, respectively. The third servo is able to bend the robot in the middle.

- Conditional operators (IF, IFLTE) are not necessary to evolve good programs and do not increase the quality significantly. On the contrary, table 4 suggests that they inhibit the evolutionary process.

- There is no increase in quality of the best individual when going from $M = 100$ to $M = 500$ (only 3 experiments with $M = 500$ have been performed).

- In successful experiments (e.g. 83 and 84), the program length increases (Figure 4).

Figure 5 shows the average forward movement of the robot during a very interesting run. In this run all six servos are involved into the movement until at time step 3300, an important joint (the clevis of joint 0, Fig. 1) between two servos broke accidently. Joint 0 connects servo 3 with servo 4 and is the only connection to the rudder horn of servo 3. The forward movement decreased drastically. However, after a short while (about 200 fitness tests), the GP system was able to compensate the error.

As an example, Figure 6 gives an impression of the behaviour of the following high-fitness individual that uses all 6 servos for movement:

(DIV (DIV (DELAY (SETSERVO4 (SETSERVO1 (ADD (MUL (SETSERVO5
(SETSERVO1 -103)) (DELAY (DELAY (SUB -56 (SETSERVO0 (SETSERVO5
(SETSERVO2 83)))))))) (DIV (DELAY (SETSERVO4 (SETSERVO1 (ADD (MUL
(SETSERVO5 (DELAY -40)) (DELAY -38)) (ADD (MUL -79 -8) (ADD
(MUL (SETSERVO5 (SETSERVO1 -103)) (DELAY (DELAY (SUB 49 (DE-
LAY (SETSERVO4 (SUB 49 (SETSERVO3 (SETSERVO5 (SETSERVO2 -
36)))))))))) (DIV (DELAY (SETSERVO4 (SETSERVO1 (ADD (SETSERVO5
(SETSERVO1 -104)) (DELAY (MUL (SETSERVO5 (DELAY -40)) (DELAY
-38))))))) (SETSERVO3 (DELAY (SETSERVO3 -38)))))))))) (SETSERVO3
(SETSERVO3 (SETSERVO0 (SETSERVO4 (DELAY (DIV (DELAY (SETSERVO4
(SETSERVO1 (ADD (MUL (SETSERVO5 (SETSERVO1 -103)) (DELAY (DE-
LAY (SUB 49 (SETSERVO0 (SETSERVO5 (SETSERVO1 -100))))))) (DIV (DE-
LAY (SETSERVO4 (SETSERVO1 (ADD (MUL (SETSERVO5 (SETSERVO1 -
104)) (DELAY -38)) (DELAY (DELAY (SETSERVO4 -37))))))) (SETSERVO3
(DELAY (SETSERVO3 -38))))))) (SETSERVO3 (DELAY (SETSERVO3 -
40))))))))))))) (ADD (MUL (DELAY (SETSERVO4 -37)) (DELAY (DE-
LAY (SUB 49 (SETSERVO0 (SETSERVO0 (SETSERVO5 (DELAY (DELAY
(SETSERVO1 -104)))))))))) (DIV (DELAY (SETSERVO5 (SETSERVO1 (ADD
(DIV (SETSERVO5 (SETSERVO3 -104)) (DELAY -38)) (DELAY (SUB 105
126)))))) (SETSERVO3 (DELAY (SETSERVO3 -38)))))) (SETSERVO3 (DELAY
(SETSERVO3 -40))))

Best individual of experiment 85 in praefix notation (ID: 1320/85)

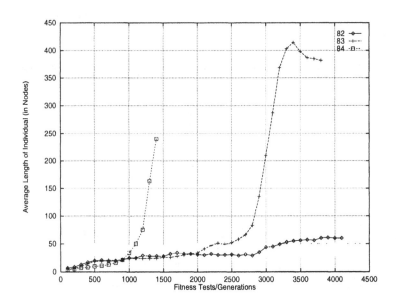

Fig. 4. Average individual length over time of three typical experiments. The successful experiments (83, 84) show a drastic increase of program length compared to
unsuccessful run 82.

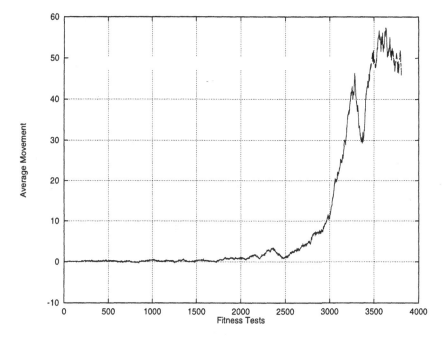

Fig. 5. Adaptation to an unexpected error (experiment 83). The average moved distance (running average over 100 fitness evaluations) is shown over time. The sudden decrease at time 3300 results from a break of an important joint connecting to servos. The increase shows that the system is able to adapt to the new situation.

4   Summary and Conclusion

An easy-to-maintain and cheap robot architecture has been presented, which has been shown to be useful as a platform for testing and demonstrating learning techniques for bio-inspired robots. It has also been shown that Genetic Programming can be used to evolve control programs in real time for an architecture for that no model exists.

The algorithm presented here needs only a very low amount of computational resources because the fitness evaluation that is performed by the robot takes most of the time. The huge amount of remaining processor time can be used to speed up the learning process, e.g. by learning from the past sensory data.

Another future direction will be to include sensor information. In the experiments presented here only the moved distance is measured and used as an input to the learning GP system. By adding sensors, two aspects will become important. The robot has to cope with sensor information (e.g., in order to follow a wall), so the operator set of the GP system has to be extended to process sensory data. Secondly, sensors may provide information about internal states of the robot, which may result in a faster learning process and a robust control strategy.

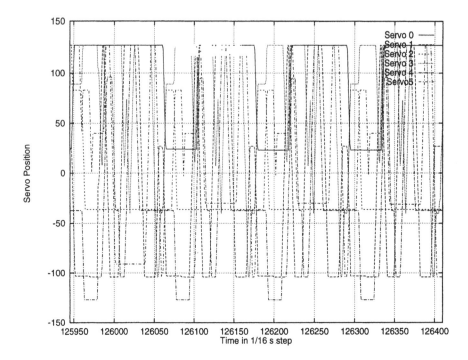

Fig. 6. Behaviour of the best individual of experiment 85. The figure should give an impression of the complex movement where all six servos are involved.

Acknowledgement

Support has been provided by the DFG (Deutsche Forschungsgemeinschaft) under grant Ba 1042/2-2. W.B. acknowledges partial support by the International Computer Science Institute, UC Berkeley, CA.

References

1. Sims, K.: Evolving 3D Morphology and Behavior by Competition, Proceedings of the 4th International Workshop on the Synthesis and Simulation of Living Systems, Artificial Life IV, pp. 28-39, MIT Press, July (1994)
2. Paap, K. L., Dehlwisch, M., Klaassen, B.: GMD-Snake: A Semi-Autonomous Snake-like Robot, In: Distributed Autonomous Robotic Systems 2, Springer-Verlag, Tokio, (1996)
3. Ostrowski J. P., Burdick, J. W.: Gait Kinematics for a Serpentine Robot, Int. Conf. on Robotics and Automation (1996)
4. Triantafyllou, M. S., Triantafyllo, G. S.: An efficient swimming machine, Scientific American, 272, pp. 64-70 (1995), see also: http://web.mit.edu/towtank/www/projects.html
5. Banzhaf, W., Nordin P., Keller, R. E., Francone F.: Genetic Programming - an Introduction Morgan Kaufmann, (1997)

6. Mataric, M. J., Cliff. D.: Challenges in evolving controllers for physical robots. *Journal of Robotics and Autonomous Systems 19*(1), 67–83 (1996).

7. Davidor, Y.: *Genetic Algorithms and Robotics.* World Scientific, Singapore, 1990.

8. Dorigo, M., Schneph, U.: Genetics-based machine learning and behavior based robotics. *IEEE Transactions on Systems, Man and Cybernetics*, 23(1), 1993.

9. Koza. J. R.: Evolution of subsumption using genetic programming. In F. J. Varela and P. Bourgine, editors, *Proceedings of the First European Conference on Artificial Life. Towards a Practice of Autonomous Systems*, pages 110–119, Paris, France, 11-13 December 1992. MIT Press.

10. Koza, J. R.: : Genetic Programming, MIT Press, Cambridge MA, (1992)

11. Lee, W.-P. Hallam, J., Lund. H. H.: Learning complex robot behaviors by evolutionary approaches. In *6th European Workshop on Learning Robots, EWLR-6*, pages 42–51, Hotel Metropole, Brighton, UK, 1-2 August 1997.

12. Nordin, P., Banzhaf, W.: Genetic programming controlling a miniature robot. In E. V. Siegel and J. R. Koza, editors, *Working Notes for the AAAI Symposium on Genetic Programming*, pages 61–67, MIT, Cambridge, MA, USA, 10–12 November 1995. AAAI.

13. Olmer, M., Banzhaf, W., Nordin. P.: Evolving real-time behavior modules for a real robot with genetic programming. In *Proceedings of the international symposium on robotics and manufacturing*, Montpellier, France, May 1996.

14. Singleton, A.: gpquick (Steady-state tree-based C++ GP-Sytem), ftp.cc.utexas.edu/pub/genetic-programming/code.

15. Salomon. R.: Scaling behavior of the evolution srategy when evolving neuronal control architectures for autonomous agents. In *Evolutionary Programming 6 6th International Conference, EP97*, pages 48–57, Indianapolis, Indiana, USA, apr 1997.

16. Steels, L.: Emergent functionality in robotic agents through on-line evolution. In Rodney A. Brooks and Pattie Maes, editors, *Proceedings of the 4th International Workshop on the Synthesis and Simulation of Living Systems ArtificialLifeIV*, pages 8–16, Cambridge, MA, USA, July 1994. MIT Press.

17. Nolfi, S., Floreano, D., Miglino, O., Mondada. F.: How to evolve autonomous robots: Different approaches in evolutionary robotics. In Rodney A. Brooks and Pattie Maes, editors, *Proceedings of the 4th International Workshop on the Synthesis and Simulation of Living Systems ArtificialLifeIV*, pages 190–197, Cambridge, MA, USA, July 1994. MIT Press.

Learning Behaviors for Environmental Modeling by Genetic Algorithm

Seiji Yamada

Department of Computational Intelligence and Systems Science
Interdisciplinary Graduate School of Science and Engineering
Tokyo Institute of Technology
4259 Nagatsuta-cho, Midori-ku, Yokohama 226-0026, JAPAN

Abstract. This paper describes an evolutionary way to lean behaviors of a mobile robot for recognizing environments. We have proposed AEM (Action-based Environment Modeling) which is an appropriate approach for a simple mobile robot to recognize environments, and made experiments using a real robot. The suitable behaviors for AEM have been described by a human designer. However the design is very difficult for them because of the huge search space. Thus we propose the evolutionary design method of such behaviors using genetic algorithm and make experiments in which a robot recognizes the environments with different structures. As results, we found out that the evolutionary approach is promising to automatically acquire behaviors for AEM.

1 Introduction

Primary research on an autonomous agent which recognizes a real environment have been done in robotics. The most studies have tried to build a precise geometric map using a robot with high-sensitive and global sensors like vision sensors [3]. Since their main aim is to navigate a robot with accuracy, the precise map is necessary. However, to recognize environments, such a strict map may be unnecessary. Actually many natural agents like animals seem to recognize environments only with low-sensitive and local sensors like touch sensors, and a precise geometric map is not necessary. In terms of engineering, it is important to build a mobile robot which can recognize environments only with the least sensors.

Thus we have tried to build a mobile robot which recognizes an environment only with low-sensitive and local sensors. Since such a robot does not know its position in an environment, it cannot build the global map of the environment using sensor data. Hence we proposed approach that a mobile robot can recognize an environment using *action sequences* generated by acting there. We call the approach AEM (Action-based Environment Modeling), and implemented it on a real mobile robot [15]. Using AEM, a robot can build a robust model of an environment only with low-sensitive and local sensors, and recognize an environment. In our research, the mobile robot is behavior-based and acts using *given* suitable behaviors (wall-following) for AEM in enclosures made of white plastic boards. Then the sequences of the actions executed in each enclosure are obtained. They are transformed into real-value vectors, and inputted to a Kohonen's self-organizing network. Learning without a teacher is done and a mobile robot becomes able to identify enclosures. We fully implemented the system on a real mobile robot with two infrared proximity sensors, and made experiments for evaluating the ability. As a result, we found out the recognition of enclosures was done well.

However, in AEM, there is a significant problem: where the suitable behaviors come from. Although the design for such behaviors is very hard because of the huge search space, it has been done by a human designer thus far. Hence we propose an evolutionary design method of suitable behaviors for AEM using GA (Genetic Algorithm), and make preliminary experiments. For future implementation on a real mobile robot, we use a Khepera simulator in the experiments. From the experimental results, we found out that the evolutionary approach is promising to automatically acquire behaviors for AEM.

In the similar approach to AEM, several studies have been done in robotics [9][11] and A-Life [12]. Nehmzow and Smithers studied on recognizing corners in simple enclosures with a self-organizing network [12]. They used direction-duration pairs, which indicate the length of walls and shapes of past corners, as an input vector to a self-organizing network. After learning, the network becomes able to identify corners. Mataric represented an environment using automaton consisting landmarks as nodes [9]. Though the representation is more robust than a geometric one, a mobile robot must segment raw data into landmarks and identify them. Nakamura et al. utilized a sequence of sonar data in order to reduce uncertainties in discriminating local structure [11]. Though the sequence consists of sensor data (not actions), their approach is similar to AEM.

Wall-following and random-walking were used as suitable behaviors in [9][12][15] and [11] respectively. The behaviors were described by a human designer, and fixed. Hence they have the same significant problem that the design of the behaviors is very difficult as AEM.

There are several studies for applying GP (Genetic Programming) [8] to behavior learning of a mobile robot [14][7][6][13]. Unfortunately, in the studies, very simple behaviors like obstacle avoidance were learned. In contrast with them, our aim is to learn the suitable behaviors to AEM, and the behaviors is complex one consisting of several kinds of primitive behaviors.

2 Action-Base Environment Modeling and Its Problem

In AEM, a mobile robot is designed with behavior-based approach [2]. The *behavior* means mapping from states to actions, and a human designer describes states, actions and behaviors so that sequences of executed actions can represent environment structure. Since AEM uses an action sequence, not sensed data, for describing an environment, the model is more abstract and robust than a geometric one [15].

An AEM procedure consists of two stages: a *training phase* and a *test phase* (Fig.1). It uses 1-Nearest Neighbor method [4], one of effective supervised-learning methods. In the training phase (Fig.1(a)), a robot is placed in a *training environments* having a *class* in which the environment should be included. Plural environments may be included in the same class. The behavior-based mobile robot acts in the environments using given behaviors, and obtains sequences of executed actions (called *action sequences*) for each of them. The action sequences (lists of symbols) are transformed into the real-valued vectors (called *environment vectors*) using chain coding [1]. The environment vectors are stored as *cases*, and the training phase finishes.

Next, in the test phase (Fig.1(b)), a robot is placed in a *test environment*: one of the training environments. The robot tries to identify the test environment with one of training environments, and we call this task *recognition of an environment*. The identification is done by determined the most similar training environment (1-Nearest Neighbor) to the test environment. The similarity is evaluated with Euclidean distance between environment vectors. The robot considers that the most similar training environment has the same class to the test environment, and recognition of environments is done.

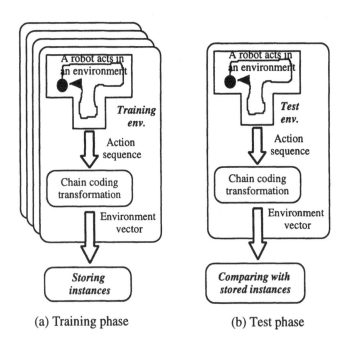

(a) Training phase   (b) Test phase

Fig. 1. Overview of AEM

However, in AEM, there is a significant problem: where the suitable behaviors come from. Since the suitable behaviors depend on an environment structure which a robot should recognize, they have been described by a human designer thus far. However the task is very difficult for him or her. Because the search space for a suitable behavior is very huge: the computational complexity is $O(a^s)$, where a and s are the number of actions and states. Thus, we propose an evolutionary method to automatically acquire such behaviors using GA.

3 Describing a Mobile Robot, States and Actions

Using real mobile robots as individuals in GA is not practical because it is impossible to operate several tens of real robots over more than one hundred generations. Thus we use a simulator for acquiring behaviors, and intend to implement the learned behavior on a real mobile robot.

3.1 A Simple Mobile Robot: Khepera

We use a miniature mobile robot Khepera™(Fig.2, the radius and the height are 25mm and 32mm) which widely used in A-Life and AI. It has Motorola 68331 Micro processor, 256KByte RAM, and is programmable. As shown in Fig.3, it also has two DC motors (two black bars in Fig.3) as actuators and eight Infra-Red proximity sensors which measure both distance to obstacles and light strength. However, since the sensor data is imprecise and local, Khepera cannot localize itself in global map. In the later experiments, the simulator build for Khepera will be used.

Fig. 2. Khepera

Fig. 3. Sensor positions

3.2 State Description

We describe a state with the range of a sensed value. For reducing the search space of behaviors, we restrict the number of states and actions as small as possible.

Though a sensor on Khepera returns 10 bit $(0 \sim 1023)$ value for distance and light strength, the value is very noisy and crisp. Thus we transform the distance value into binary vlues 0 and 1. The value "0" means an obstacle exists within 3cm from a robot. The value "1" means it does not exist. Furthermore only three (front, left and right) of eight sensors are used for reducing states.

Next states for light strength are described. Since only simple behaviors like approaching, leaving a light are considered suitable to AEM, we describe the state using the label of the sensor with the strongest light value and binary values which mean a light is "near" or "far". As well as states for distance, the only 4 sensors (front, left, right and back) are used. Additionally a state in which all of the sensors has almost same values is also considered. As a result, the number of states fot light is nine ($= 2 \times 4 + 1$). The total number of states is 72 ($= 2^3 + 9$)

3.3 Action Description

The following four actions are described. In experiments in the past research [15], we found the actions is sufficient for a mobile robot to do simple behaviors like wall-following. A mobile robot acts in an environment by executing the actions, and consequently an action-sequence like $[A2, A4, A1, A1, \cdots]$ is obtained.

- A1: Go 5mm straight on.
- A2: Turn 30° left.
- A3: Turn 30° right.
- A4: Turn 180°.

3.4 Transformation into an Environment Vector

The generated action-sequence is transformed into an environment vector. Let an action-sequence and its environment vector be $[a_1, a_2, \cdots, a_n]$ ($a_0 = 0$, $a_i \in$ {A1, A2, A3, A4}) and $V = (v_1, v_2, \cdots, v_m)$ respectively. The vector values of V are determined by the following rules. They change the vector value when the direction of movement changes. These rules are considered a kind of chain coding [1]. For example, an action sequence $[A2, A2, A3, A3, A3, A4, A1]$ is transformed into an environment vector $(1, 2, 1, 0, -1, 1, 1)$.

1. If $a_i = A1$ then $v_i = v_{i-1}$.
2. If $a_i = A2$ then $v_i = v_{i-1} + 1$.
3. If $a_i = A3$ then $v_i = v_{i-1} - 1$.
4. If $a_i = A4$ then $v_i = -v_{i-1}$.

As mentioned in §2, in training phase, training environments are given to a robot for learning. The robots acts in the given environments, and stores the environment vectors transformed from the action sequences. Next, in test phase, test environments are given to the robot. It identifies the test environment with one of training environments by 1-Nearest Neighbor method using Euclidean distance as similarity.

4 Applying Genetic Algorithm to Acquire Behaviors

Since the number of states is 72 and that of actions is four, the number of possible behaviors is $4^{72} = 2.23 \times 10^{43}$. We have to search the suitable behaviors to AEM in such a huge search space. Genetic algorithm [5] is applied to the search because it does not need any domain-dependent heuristics.

4.1 GA Procedure and Coding Behaviors

In the followings, we describe GA procedure used in our research.

Step1 *Initializing population*: An initial population $I_1, \cdots, I_N$ are randomly generated.

Step2 *Computing fitness*: Compute the fitness $f_1, \cdots, f_N$ for each individual $I_1, \cdots, I_N$.

Step3 If a terminal condition is satisfied, this procedure finishes.

Step4 *Selection*: Using $f_1, \cdots, f_N$, select a child population C from the population.

Step5 *Crossover*: Select pairs randomly from C on probability P_{cross} (called *crossover rate*). Generate two children by applying crossover to each pair, and exchange the children with the pairs in C.

Step6 *Mutation*: Mutate the individuals in C based on mutation rate P_{mut}.

Step7 Go to **Step2**.

We set the following parameters which are considered effective experimentally.

- *Population size*: 50
- *Selection method*: Elite strategy and tournament selection (the size = 2).
- *Crossover operator*: Uniform crossover.
- *Crossover rate P_{cross}*: 0.8
- *Mutation rate P_{mut} per gene*: 0.05

Since we deal with deterministic action selection, not probabilistic, the behavior is mapping from a single state to a single action. Thus we use the coding in which one of actions $\{A1, \cdots, A4\}$ is assigned to each state of $s_1, \cdots, s_{72}$ (Fig.4).

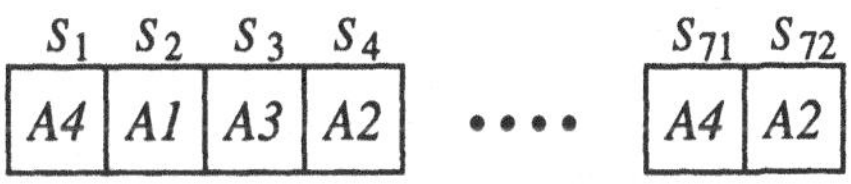

Fig. 4. A coded behavior

4.2 Defining a Fitness Function

Fitness is a very important for GA. Thus we have to carefully define the fitness function for AEM. We consider three conditions for suitable behaviors to AEM: *termination of actions, accuracy of recognition* and *efficiency of recognition*. The fitness functions for each conditions are defined, and then are integrated.

Termination of Actions A mobile robot needs to stop its actions by itself. Otherwise it may act forever in an environment, and no action sequence is obtained. Thus the termination of actions is the most important condition. We use *homing* which a robot returns his home. Because homing is considered a general method to terminate actions, and makes the length of an action sequence depend on the size of an environment. A method to terminate actions within a fixed length of an action sequence does not have such advantage. The homing concretely means turning to the neighborhood of a start point, and the termination is evaluated with the following function g. Its range is $[0, 1]$, and returns 1 when a robot succeeded in homing in all the training environments.

$$g = \frac{\text{(No. of E-trial)} + \text{(No. of H-trial)}}{2 \times \text{(Total No. of trials)}}$$

where E-trial and H-trial means trials in which a robot escaped from the neighborhood of the start point and trials in which it succeeded in homing.

Accuracy of Recognition Another important criterion is accuracy of identifying test environments. The accuracy is evaluated with the following function h. Its range is $[0, 1]$, and when $h = 1$, all the test environments were recognized correctly.

$$h = \begin{cases} \dfrac{\text{No. of correctly identified test env.}}{\text{Total No. of test env.}} & g = 1 \\ 0 & 0 \leq g < 1 \end{cases}$$

Efficiency of Recognition In AEM, a robot needs to *act* by operating actuators for recognizing an environment, and the actions significantly cost. Hence the actions should be as small as possible for efficiency of recognition. We hence introduce the following fitness function for evaluating the efficiency[1].

$$k = \begin{cases} 1 - \dfrac{\sum_{i=1}^{n} S_i}{n * S_{max}} & h = 1 \\ 0 & 0 \leq h < 1 \end{cases}$$

where S_i is the size of an action sequence obtained in ith test environment, S_{max} is the limited size of an action sequence, and n is the number of test environments. The function have range $(0, 1]$ and has larger value as more efficient.

---

[1] The obstacle avoidance is implicitly evaluated by the function k because the collision increases the length of an action sequence.

We finally integrate three fitness function into the following fitness function f having range $[0, 3]$, and it is used in this research. Since the function h (or k) takes 0 when g (or h) does not take 1, the function f is phased: the termination of actions is satisfied when $1 \leq f$, the recognition is completely correct when $2 \leq f$, and the recognition efficiency is improved when $2 \leq f < 3$.

$$f = g + h + k$$

5 Experiments with Simulation

It is impractical that we use real robots as individuals in GA. Thus we implement the system using a Khepera simulator [10], and make experiments in it. The parameters used in all experiments are described in the followings: the neighborhood of a start point is a circle with 100mm radius, and the limited size of an action sequence is 2000 actions.

In the simulator, the motor has $\pm 10\%$ random noise in velocity, $\pm 5\%$ one in rotation of robot's body. Furthermore an Infra-Red proximity sensor has $\pm 10\%$ random noise in distance, and $\pm 5\%$ one in light strength. These noise makes the simulator close to a real environment.

If a robot cannot return home within the limited size, the trial ends in failure. If the fitness value becomes more than two, the trial ends in success, and then both termination and accuracy are satisfied. In all the following experimental results, we show one of success trials, not averaged results.

In all experiments, we give each of training environments to a robot once. The robot acts in the environments, and the environment vectors transformed from the action sequences are stored as instances. Next each of the *training* environment is given to the robot as a *test* environment, and the robot identifies each of the test environments with one of the training environments. The start points and directions is fixed in bottom center and left.

Note that though the test environments are same to training environments, the action sequences are different because of the random noise in a simulator. Thus the obtained behavior by our method has robustness[14][6].

In all the experiments, we had 20 trials which have different initial conditions for GA and the generation was limited to 100. Some trials failed depending on the initial condition. In the followings, we present the succeeded trials for each experiment.

5.1 Exp-1: Environments with Different Contours in Shape

First we made experiments Exp-1 using environments with different contours in shape. Parts of five environments: { empty, L, L2, invL, small-empty } are given to a robot, and we investigated the ability to recognize them. Additionally twelve different shape environments including the four ones are used.

The experimental results are shown in Table.1, and Fig.5 indicates the trace of the robot with the maximum fitness in the expriment for the five environments.

187

The "GN" is the generation number in which the fitness value becomes more than two, and "Max fitness" means the maximum fitness value.

In such simple environments, the suitable behaviors for AEM were obtained within few generations. Seeing from Fig.5, different action sequences were obtained depending on the environment structure.

Table 1. Experimental results in Exp-1

Training environments	GN	Max fitness
(1) { empty, L }	1	2.84
(2) (1) + L2	2	2.40
(3) (2) + invL	3	2.44
(3) + small-empty	2	2.46
twelve environments	9	2.45

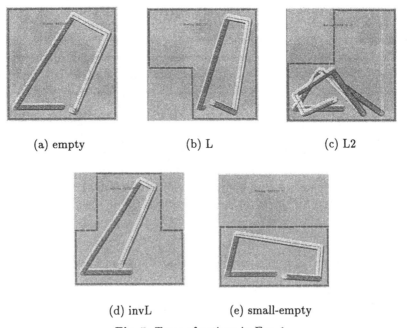

(a) empty   (b) L   (c) L2

(d) invL   (e) small-empty

Fig. 5. Trace of actions in Exp-1

5.2 Exp-2: Environments with Different Lights in Number and Position

Next, by adding different lights to environments in number and position, we made five environments: { empty, 1-lamp, 2-lamp, 3-lamp, 4-lamp }. Exp-2 is made by using parts of the environments. A light was so strong that a robot can detect the light direction in any place. The experimental results are shown in Table.2. Fig.6 indicates the trace of the robot with the maximum fitness in the experiment for the five envirnments. In the figures, a black circle stands for a light.

Though the GN increased more than ones in Exp-1, the suitable behaviors were obtained. In Fig.6(a) ~ Fig.6(c), the actions are slightly different. In contrast with them, the actions of Fig.6(d) and Fig.6(e) are significantly different from that of Fig.6(a) ~ Fig.6(c). The lights in the left area seem to influence them strongly.

Table 2. Experimental results in Exp-2

Training environments		GN	Max fitness
(1)	{ empty, 1-lamp }	1	2.85
(2)	(1) + 2-lamp	7	2.53
(3)	(2) + 3-lamp	8	2.77
	(3) + 4-lamp	12	2.77

5.3 Exp-3: Environments with Different Contours and Lights

We set six environments: { empty, 1-lamp, L, L-1-lamp, invT, invT-1-lamp } by adding a light to three environments with different contours, and made experiments Exp-3 using them. Each environment is included different class and all of them should be distinguished.

As a result, GN was 13 and the maximum fitness was 2.38. Fig.7 shows the trace of the robot with the maximum fitness. Over all the environments, actions are very different mutually.

5.4 Exp-4: A Single Class Includes the Plural Training Environments

In Exp-1 ~ Exp-3, all the environments are included in different classes. However, in Exp-4, plural environments are included in a single class. For recognizing such environments, generalization is necessary. We assigned three classes to six environments used in Exp-3: { empty, 1-lamp }, { L, L-1-lamp }, { invT, invT-1-lamp }, and made experiments. As a result, GN was 15 and the maximum

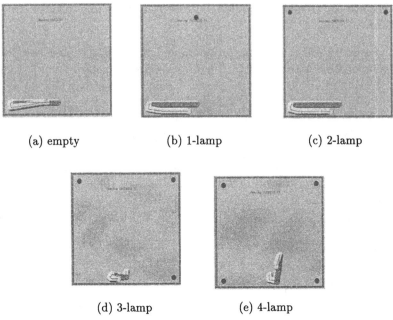

(a) empty   (b) 1-lamp   (c) 2-lamp

(d) 3-lamp   (e) 4-lamp

Fig. 6. Trace of actions in Exp-2

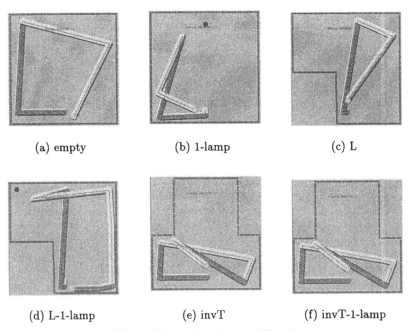

(a) empty   (b) 1-lamp   (c) L

(d) L-1-lamp   (e) invT   (f) invT-1-lamp

Fig. 7. Trace of actions in Exp-3

fitness was 2.63. Thus our approach is valid for induction from sevral instances of a class.

6 Conclusions

We proposed the evolutionary acquisition of suitable behaviors to Action-based Environment Modeling. GA was applied to search the behaviors, and the simulated mobile robots were used as the population. States and actions were described for coding chromosomes. We made experiments in different environments in shape and lights, and found out our approach is effective to learn behaviors. However there are open problems like the followings.

- *Analysis and more complex domain*: We must analyze the experimental results for clarify how to acquire the suitable behaviors. Furthermore we will make experiments in more complex domains, and clarify problems there.
- *Robustness against initial conditions*: In all the experiments, start points and start direction were fixed. However, when a real robot is used, such a precise initial situation are impractical. Thus we must attempt experiments in which the initial situation is noisy. The learning may be difficult because a mobile robot acts sensitively to the initial situation [14].
- *Implementing learned behaviors on a real robot*: Implementation on a real robot is our final target. The gap between simulation and an real environment may make it difficult.

References

1. E. M. Arkin, L. P. Chew, D. P. Huttenlocher, K. Kedem, and J. S. B. Mitchell. An efficiently computable metric for comparing polygonal shapes. *IEEE Transaction on Pattern Analysis and Machine Intelligence*, 13(3):209–216, 1991.
2. R. A. Brooks. A robust layered control system for a mobile robot. *IEEE Transaction on Robotics and Automation*, 2(1):14–23, 1986.
3. J. L. Crowly. Navigation of an intelligent mobile robot. *IEEE Transaction on Robotics and Automation*, 1(1):31–41, 1985.
4. B. V. Dasarathy. *Nearest Neighbor (NN) Norms: NN Pattern Classification Techniques*. IEEE Computer Society Press, 1991.
5. J. Holland. *Adaptation in Natural and Artificial Systems*. University of Michigan Press, 1975.
6. T. Ito, H. Iba, and M. Kimura. Robustness of robot programs generated by genetic programming. In *Genetic Programming 1996, Proceedings of the First Annual Conference*, pages 321–326, 1996.
7. J. R. Koza. Evolution of subsumpton using genetic programming. In *Proceedings of the First European Conference on Artificial Life*, pages 110–119, 1991.
8. J. R. Koza. *Genetic Programming*. MIT Press, 1992.
9. Maja J. Mataric. Integration of representation into goal-driven behavior-based robot. *IEEE Transaction on Robotics and Automation*, 8(3):14–23, 1992.
10. O. Michel. *Khepera Simulator v.2 User Manual*. University of Nice-Sophia Antipolis, 1996. (http://wwwi3s.unice.fr/~om/khep-sim.html).

11. T. Nakamura, S. Takamura, and M. asada. Behavior-based map representation for a sonor-based mobile robot by statistical methods. In *1996 IEEE/RSJ International Conference on Intelligent Robots and Systems*, pages 276–283, 1996.

12. U. Nehmzow and T. Smithers. Map-building using self-organizing networks in really useful robots. In *Proceedings of the First International Conference on Simulation of Adaptive Behavior*, pages 152–159, 1991.

13. P. Nordin and W. Banzhaf. A genetic programming system learning obstacle avoiding behavior and controlling a miniature robot in real time. Technical report, Department of Computer Science, University of Dortmund, 1995.

14. C. W. Reynolds. Evolution of obstacle avoidance behavior: Using noise to promote robust solutions. In Jr. K. E. Kinnear, editor, *Advances in Genetic Programming*, volume 1, chapter 10, pages 221–241. MIT Press, 1994.

15. S. Yamada and M. Murota. Applying self-organizing networks to recognizing rooms with behavior sequences of a mobile robot. In *IEEE International Conference on Neural Networks*, pages 1790–1794, 1996.

Evolving and Breeding Robots

Henrik Hautop Lund   Orazio Miglino

The Danish National Centre for IT Research – CIT
University of Aarhus, Ny Munkegade, bldg. 540, 8000 Aarhus C., Denmark
hhl@daimi.aau.dk
http://www.daimi.aau.dk/~hhl/

Faculty of Psychology
II University of Naples, Viale Giovanni Paolo I, 81055 S. Maria C. V. (CE), Italy
orazio@caio.irmkant.rm.cnr.it
http://gracco.irmkant.rm.cnr.it/orazio/

Abstract Our experiences with a range of evolutionary robotic experiments have resulted in major changes to our set-up of artificial life experiments and our interpretation of observed phenomena. Initially, we investigated simulation-reality relationships in order to transfer our artificial life simulation work with evolution of neural network agents to real robots. This is a difficult task, but can, in a lot of cases, be solved with a carefully built simulator. By being able to evolve control mechanisms for physical robots, we were able to study biological hypotheses about animal behaviours by using exactly the same experimental set-ups as were used in the animal behavioural experiments. Evolutionary robotic experiments with rats open field box experiments and chick detours show how evolutionary robotics can be a powerful biological tool, and they also suggest that incremental learning might be fruitful for achieving complex robot behaviour in an evolutionary context. However, it is not enough to evolve controllers alone, and we argue that robot body plans and controllers should co-evolve, which leads to an alternative form of evolvable hardware. By combining all these experiences, we reach *breeding robotics*. Here, children can, as breeders, evolve e.g. LEGO robots through an interactive genetic algorithm in order to achieve desired behaviours, and then download the evolved behaviours to the physical (LEGO) robots.

1 Introduction

In the field of evolutionary robotics, many researchers have tried to transfer a corpus of knowledge from previous experiences in evolving simulated artificial agents (e.g. controlled by neural networks [18, 24]) to model psychobiological phenomena. However, there are some difficulties that one has to consider. The first problem that researchers in this field had to face was how to avoid the extremely time consuming cost of genetic algorithms directly applied to real robots. Consider the following example, taken from some of our earlier experiments. In order to test a single mobile robot, it is allowed to run for 500 actions in an environment. In order to account for initial starting position biases, the mobile robot is run for 3 epochs with different starting positions. Since each action takes 100 m sec. to perform, the testing of the single robot's 1500 actions will take 150 sec. The evolution process of a population of 100 control systems for 100 generations will therefore take a minimum of 1,500,000 sec. or approximately 17 days (not accounting for the time spent on reproduction and movement of the robot to new initial positions). To report scientifically valid results, we would like to do the single test 5–10 times with different initial random seeds. This means that the robot has to run for half a year for us to obtain valid results of just one experiment! Such a time consumption makes the approach infeasible, so researchers have tried to reduce it. We [16, 22] proposed to make part of evolutionary process in a simulated environment and a part in the physical environment. This approach required to: (a) specify a methodology for building an efficient simulator of the physical characteristics of robot and environment; (b) develop genotypes and techniques that produce plastic phenotypes that absorb the change of environments (e.g. from simulated to real environment). In section 2, we report our experiences with these problems.

However, after having built a robust setting to produce evolutionary robotic experiments, we had to confront the efficiency of the techniques to develop more complex behaviours, processes and architectures. In particular, section 3 reports some of our experiences (a) when simulating the behaviour of rats in a classical experimental setting, (b) simulating chicken detour behaviour, (c) evolving the hardware body structure. Thanks to a LEGO research and development contract, we recently explored the possibility of using evolutionary robotics in order to build an adaptive artificial pet as a toy for children. This experience (described briefly in section 4) represents the ultimate edge of our common work and convinces us that evolutionary robotics is rapidly becoming a sort of "Breeding Robotics". Our attempts to produce complex and adaptive behaviours made us orient the evolutionary process so that the pre-designing of environments, neural architectures, and fitness functions prompted the emergence of more sophisticated behavioural strategies. In substance, we coupled the power of an evolutionary process with the informality of a "breeder of robots".

2 Evolutionary Robotics Set-Up

2.1 Evolving On-line or in Simulators?

As reported above, the on-line approach to evolutionary robotics means that the robot has to run for at least half a year for us to obtain valid results of just one experiment. Such a time consumption makes the approach infeasible, so researchers have tried to reduce it. One way to do so is to use a shorter time slice than 100 m sec. per action. This requires that the control system is not too complex, so that the appropriate motor responses can be calculated within this time slice. Another way is to test the robot for fewer actions — for example, testing the robot for only 200 actions for only 1 epoch will reduce the time consumption of a single evolutionary process in the example given above from 17 days to less than 2.5 days. This is the approach taken by researchers who do the evolution process entirely on the real robots (e.g. [4]). The approach has the disadvantage of not allowing many testings of each single control system (few actions and only 1 epoch), so the performance of a single control system will be heavily biased by the initial starting position of the robot.

Because of the problems with on-line evolution mentioned above, we have advocated a scheme to build simulators for robots, and then evolve the control systems in simulation before transferring the evolved control systems to the real robots in the real environments. Other researchers [10] have suggested using a mathematical description of motor and sensor responses of the specific robot. Their simulator is based on a set of equations, that should model the real world physics, which are used by the simulator to calculate specific values for the IR sensors and/or ambient light sensors and wheel speeds for each action of a specific robot. It is assumed that the responses of all sensors of a robot have the same characteristics, which is unfortunately not so! The individual sensors vary quite a lot from each other in identical conditions, so an accurate simulator must take the idiosyncrasies of each sensor into account. Despite the inaccuracy of a mathematical model, it can be used effectively for experiments where high accuracy of transition of control systems from simulation to reality plays a minor role in the first instance. We have used such a mathematical model, where motor responses are decided by differential equations, in our studies of the evolution of robot body plans. In that study, the immediate goal is not to be able to transfer each single control system and robot body plan to a real robot, but in the end simply that of choosing the best controller plus body plan to be transferred to a robot built according to the evolved parameters.

Another methodology to build the simulator for a robot and its environment, is to build look-up tables of the robot's sensor and motor responses environments, as we proposed. By sampling the real world through the sensors and the actuators of the robot itself, one can build a quite accurate model of a particular robot-environment dynamics, and by using look-up tables constructed by this sampling instead of mathematical models of the sensor and motor responses, one obtains both very high performance when transferring the evolved control systems from simulation to the real robot, and a huge reduction in time consumption. The on-line approach takes 625 times as long, and the mathematical model simulator approach takes 3 times as long as the run with our look-up table approach.

The disadvantage of using a simulator based on the robot's own samplings of sensor and motor responses is that these responses have to be recorded. This can be done almost automatically with some robots in some simple environments, but it still has to be solved how to do this in more complex environments, where the construction of the look-up tables might prove more time expensive. It is simple to measure responses from sensors such as bumpers and black/white detectors, while it is more tricky for sensors such as infra-red sensors and cameras. Since each sensor is not placed in a pre-defined position on e.g. a LEGO robot, we have to run sampling procedures for each single sensor consecutively. When the positions of the sensors on the robot are pre-defined, e.g. on the Khepera robot and most other robots, the sampling can be done in parallel by taking samples from all sensors at the same time, but the sampling procedure has to be sequential when this is not the case. After taking samples from specific angles and distances with each of the robot's sensors, we can combine the look-up table with a kind of mathematical approach in which a mathematical description is used to calculate the entry in the look-up table for a specific position of the robot and a specific placement of the sensors on the robot. Further, to allow for non-symmetrical objects in an environment, samples must be taken with different angles of the different object-types towards each sensor. Therefore, the sampling procedure can be time-consuming in such cases.

The advantage of using this approach to build the simulator is that the sensor and motor responses are recorded by the robot itself. The look-up tables represent how the robot itself senses and moves in the environment, and not how we, as external observers, believe that the robot interacts with the real world. There is no need for symbolic or mathematical descriptions of objects or robot responses in this methodology. Therefore, the simulator becomes a very precise model of the real world situation, and the evolved control systems can be transferred successfully from the simulator to the real robot that interacts in the real world.

2.2 Evolving Neurocontrollers

In our first evolutionary robotics experiments [16, 22, 23], we were interested in investigating how to construct neural network control systems and how to develop these. We found that the most simple solution often turned out to be the best, namely in order to construct a neural network control system for a robot, one can simply use the robot's sensory activation as input stimuli for the neural network, and allow the output activation of the neural network to control the motors of the robot. To generate a simple feed-forward neural network control system for the Khepera robot, one can connect the sensors with the motors, eventually with an intermediate layer of hidden units.

To simulate the evolution process that should develop the neural network control system for the Khepera robot, here, our version of the evolutionary robotics technique used a genetic algorithm (in other experiments, we used genetic programming). In our first experiments, we looked at the task of having the Khepera robot to perform an obstacle avoidance behaviour by moving forward as fast as possible, moving in as straight a line as possible and keeping as far away from objects as possible – a task that has now become a standard task in the field. In order to evaluate individual performance we used equation 1

$$F = \sum_{i=1}^{500} V_i(1 - DV_i^2)(1 - I_i) \tag{1}$$

where V_i is the average rotation speed of the two wheels, DV_i is the algebraic difference between signed speed values of the wheels, and I_i is the activation values of the proximity sensor with the highest activity at time i.

Since we wanted to address the problem of developing controllers within a reasonable time, we found it infeasible to run the whole evolution process on-line, and adopted a simulated/physical approach, where most of the evolution process ran in simulation.

For the obstacle avoidance task, we constructed the real, physical environment as a 60*35 cm rectangular arena surrounded by walls with 3 round obstacles of 5.5 cm placed in the centre. Walls and obstacles were covered with white paper. In order to construct a model of this environment, the Khepera robot empirically sampled the different classes of objects in the environment (wall and obstacles) through its own real sensors. The Khepera robot turned 360 degrees at different distances with respect to a wall and to an obstacle, while, in the meantime, recording the activation level of the sensors. The resulting matrices were then used by the simulator to set the activation levels of the simulated robot depending on its current position in the simulated environment. In the same way, to model the robot's motors, the effect of the different motors settings in the real world was sampled. For all possible states of the motors, it was modelled how the robot moved and turned. The obtained measures were used by the simulator to set the activation level of the neural network input units, and to compute the displacement of the robot in the simulated environment.

The procedure of letting the robot itself construct the model to be used by the simulator has several advantages: it is simple and it accounts for the idiosyncrasies of each individual sensor. It allows one to build a model of an individual robot taking into account the specificities of that robot that makes it different from other apparently identical exemplars. It also accounts for the idiosyncrasies of the environment by empirically modelling the environment itself, instead of building a mathematical model of it. We may however encounter problems such as objects of the same types may be perceived differently because of variation in the ambient light or because of slight differences in the objects themselves, actuators may produce unpredictable, uncertain effects, and the ground may present irregularities. Consequently, one would expect a decrease in performance when transferring a robot control system developed in a simulator to the real robot in the real world. Indeed, under "normal" conditions, a decrease in performance takes place when the control systems evolved in the simulator are transferred to the real robots in the real environment [22]. This is because of difficulties in constructing a simulator that is accurate enough to capture all the features of the real world that is important to the robot. To avoid this problem, noise can be applied in the simulated environment both to the simulated robots perception of objects and to the movement of the simulated robot. The experiments with noise added to the simulator show a perfect mapping in performance from the simulated to the real environment (Fig. 2(a)). If, for instance, the evolutionary development of the neural network control system for the Khepera robot takes place for 200 generations in the simulator, and the neural network control system is transferred to the real Khepera robot in the real environment, then there will be no decrease in fitness score (performance) when the control system is transferred. Further, if the evolution continues for 20 generations in the real Khepera robot in the real environment, then there will be a slight increment in fitness score.

A two factor ANOVA (Fig. 1) applied to performances obtained by individuals at the transferring moment in the simulated and the real environment shows: (a) a statistical significance of transferring the control systems from the simulator with no noise to the real environment, and (b) no statistical significant effect of transferring the control systems from the simulator with noise to the real environment. This result supports the claim that the decrease in performance can be avoided by adding noise to the simulator. In metaphoric terms, adding noise permits the emergence of artificial "genotypes" that adapt to different environments.

It is remarkable, that at the level of behaviours, the simulated and the real robots

	Degrees of freedom	Ratio of Fisher	Interval of confidence	Remarks
No noise	1/998	164.14	0.001	Significant
Conservative noise	1/998	0.74	0.001	Not significant

Figure1. Results of two factors analysis of variance between performance of individuals at generation 200 in the simulated and in the real environment.

perform identically (Fig. 2(b)). The trajectories of the paths for the simulated and the real robots match almost perfectly. In this way, the evolutionary robotic technique suggests improvement of solutions to some old, well-known problems of robotics: how to construct an accurate model of the real environment, and how to develop control systems without explicitly designing submodules of the control system.

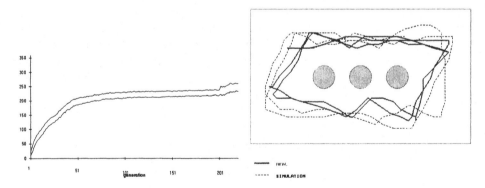

Figure2. (a) Peak and Average performance as an average result of 5 simulations. The first 200 generations are evolved in the simulated environment, the last 20 generations in the real environment. (b) Trajectory of the best individuals of generation 200 of a simulation in the simulated (dashed line) and real (full line) environment. © Miglino, Lund and Nolfi, 1995.

3 Complex Behaviours

Following our initial evolutionary robotics experiments that showed that it was indeed possible to evolve neural network controllers in simulation and then transfer them to the real robot in the real environment with no loss in performance, we naturally wanted to try to evolve more complex robot behaviours. First, we wanted to show how evolutionary robotics can be used as a tool in behavioural biology in order to verify/falsify hypotheses about animal control mechanisms. The examples that we used were the open field box experiments with rats and detour experiments with chicks.

3.1 Using Evolutionary Robotics to Test Conclusions of Experiments with Rats

During the last decade, researchers have used open field box experiments with rats in order to scientifically show that rats form cognitive maps to be used behaviourally in navigation tasks [1, 5, 20]. In open field box experiments, rats that have been shown the location of hidden food in a rectangular box are able to navigate towards and dig at that location (or at the rotational equivalent location) in a second identical box. Results like this lead researchers to state that rats use a cognitive map of the environmental shape. We studied a mobile robot in a similar experimental settings. Our "artificial organism" (i.e. robot) had no capability of constructing internal maps of the surrounding environment's geometrical shape. The robot, despite its sensory-motor limitations, performed the open field box experiments as well as rats. Our results show that the open field box experiments are not sufficient to conclude a construction of cognitive maps by the rats. Other task solutions based on no explicit knowledge of the environment are theoretically possible. In fact the geometrical properties of the environment can be assimilated in the sensory-motor schema of the robot behaviour without any explicit representation. In general, our work, in contrast with traditional cognitive models, shows how environmental knowledge can be reached withthout any form of direct representation (cognitive maps).

More precisely, what we did in these experiments, was to accurately replicate the animal behavioural experiments. A rectangular box was made out of wood and covered with white paper. As in the Gallistel experiments [5, 20], it measured 120*60 cm and was divided into 100 12*6 cm rectangles with lines on the floor (nine lines parallel to the box's short walls and nine lines parallel to the box's long walls). Likewise, the target area was a circular zone with a radius of 15 cm. A simple perceptron control system was constructed for the Khepera robot. By connecting the input of the 8 sensors to the 2 motors with traditional neural network connection weights, we obtain a simple perceptron (with linear output units). This simple control system does not have the capability of building an internal map of the surrounding environment, but it can simply make some sensory-motor responses. Again, we used a genetic algorithm to develop the controllers, as described above. The best performing neural network of the last generation (generation 30) was considered the final control system for the robot and, as in Gallistel's experiments, it was tested from 8 different starting positions: either at the middle of a wall, facing the box's centre or at the centre, facing the middle of a wall. The learning/testing process was repeated for each possible position of the target among the 80 intersections of the lines dividing the floor when excluding the centre intersection (point 5,5).

Even with as simple a control system as the perceptron for the Khepera, the robot is able to navigate to the target in the rectangular box. The table in figure 3 shows that as is the case with rats, the robot will however navigate to the rotational equivalent area as many times as to the correct target area. In 82% of the trials, the robots navigated to locations with the correct geometric relationship to the space defined by the shape of the box, but half of the times it was at the rotational equivalent of the target area. The number of successes are comparable to the ones obtained with rats, and the robots perform less misses.

In the experiment with rats, the authors reported only the results on behavioural

	% Correct	% Rotational Errors	% Misses
Rats	35	31	33
Robots	41	41	18

Figure3. Percent of navigation to Correct Area, Rotationally Equivalent Area, and Incorrect Locations. Data on rats taken from [20].

indexes (percentage of correct loaclization, rotational error, and misses). Perhaps, a careful analysis of behavioural trajectories could give more information about cognitive processes involved in the task solution. Figure 4 presents typical Khepera trajectories for a target area located in a corner. In general, for all starting positions, the robot touches a long side of the box. If we did not know the strong sensorial limitations of Khepera we could hypothesise that Khepera discriminates long vs. short sides. But, if we analyse figure 4 carefully, we can describe this strategy by a set of rules : 1) initially, when no sensors are active, Khepera produces a curvilinear trajectory on its left side; 2) when the robot touches the wall (the right-side sensors are active), then it turns left; and 3) Khepera goes on following the wall until it reaches a corner. Although the shape of the box (its geometrical characteristics) cannot be seen and represented by Khepera, it is assimilated in the behavioural sequences of the robot. For example, in order to touch a long side from every starting position, the radius of the initially curvilinear trajectory must be correlated with the size of the box.

These evolutionary robotics experiments contradict hypotheses based on animal behavioural experiments alone, and therefore the robotics experiments can be a powerful tool in biology. It suggests that animal psychologists should not only report behavioural indexes but also analyse, with attention, the behavioural tra-

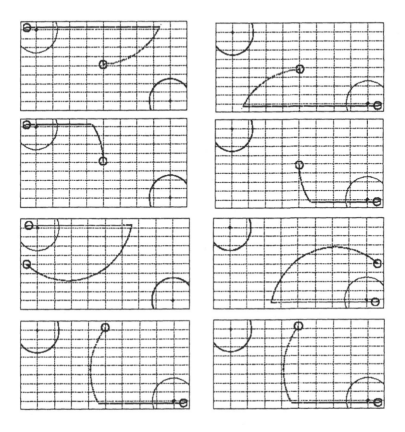

Figure4. The trajectory of the evolved Khepera robot in the open field box when started in the 8 starting positions. © Miglino and Lund , 1997.

jectories of the animals or to provide neurophysiological evidence that account for hypothesised activation of cerebral modules in such experiments.

3.2 Towards Complex Behaviours: Incremental Learning

Another example of the use of evolutionary robotics to test animal behavioural hypotheses is found in detour behaviour experiments. We will here look at this example, but put emphasis on how the robot detour experiments might help us achieving more complex behaviours with the evolutionary technique.

Animal psychology literature has intensively investigated detour behaviour in several species of mammals [12, 32]. In substance, a detour task consists in reaching a target following trajectories that, for specific environmental constraints (obstacles, different light conditions, etc.), deviate from an optimal approaching path.

One of the most used settings to observe detour behaviour is represented by a situation where the target is placed behind a U-shaped barrier (see figure 5b). In this case, in order to reach the target, an organism must make a detour around the obstacle and therefore in some input-output cycles it loses sight of the target. An organism which is able to make a detour around an obstacle is said to posses a representation of the spatial environment [28]. In fact, if the organism is able to negotiate the obstacle and reach the target, this implies that the organism is in possession of a spatial representation of the environment that tells the organism where the target is even if the target is currently not visually accessible.

Two-day-old chicks perform quite well on a detour task with a U-shaped barrier [28, 34]. Unfortunately, the complexity of natural experimental settings obliges researchers to deduce neural mechanisms and cognitive processes underlying detour abilities by just monitoring behavioural indexes (time spent in a particular area, time spent to reach a target, etc.). In fact, the question about how chicks solve the detour task is still open: "... Chicks could have used a motor algorithm with instruction such as if you turned right (left) before the goal disappeared, then turn right (left) to find it again. Dead reckoning could be a possibility: chicks may be able to continuously update their position with respect to the goal in a represented space moment by moment..." [28] pp. 4. Given this state of the art in animal spatial behaviour, we think that evolutionary robotics could be useful to give insights and computational models. In fact, by designing robotics experiments that reproduce exactly the animal experimental setting it is possible study "a posteriori" the computational structures that emerged through the learning/evolutionary process and confront them with the strategies of real organisms.

According to that perspective, we evolved different neural controllers (i.e. different neural architectures: perceptron, with hidden layer, with Elman memory units, etc.) for the Khepera robot equipped with a linear camera (for details see [21]). The environment where Khepera acted was a rectangular walled arena of 55x40 cm. It contained a low obstacle that had a U-shaped obstacle and a cylindrical target (see figure 5b). The proximity sensors of the robot can sense the obstacle or the target or the peripheral wall provided the robot is sufficiently close to them. The camera can see the target (but not the obstacle or the wall) at any distance (provided the target falls within its restricted visual field) even when the target

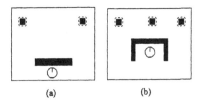

(a)   (b)

Figure5. The environments used in detour experiments: (a) in the first 200 generations; (b) after 200 generations. The dotted areas identify alternative target positions.

is on the other side of the low obstacle. At the beginning of each trial, Khepera was positioned in the environment behind the obstacle facing North. The target was positioned beyond the obstacle. The fitness function was based on two components. The first component was how much time Khepera spent in each trial near the target, where nearness to the target means to be located 12 cm or less from the target. The second component is calculated according to formula (1) mentioned in section 2.2, however where only three IR sensors where used. In other words, the second component of fitness rewards the individual, in each step, if the individual moves fast and straight (without turning) and it punishes the individual for being too close to the obstacle or wall (i.e., when the infrared sensors are activated; notice that the second fitness component is ignored when the robot is near the target).

The situation described above reflects the characteristics of the detour task used in animal experiments. In fact, the only solution for the robot is to turn right or left and make a detour around the wall in order to reach the target. But this creates a serious problem for our robot. By turning to its right or left, the robot is turning away from the target and therefore the target is likely to leave the robot's visual field. The problem is particularly serious if the obstacle has the shape shown in figure 5. In order to negotiate the obstacle, the robot may be forced to go in the opposite direction with respect to the target and therefore to loose sight of the target.

After many experiments, we reached a corpus of data that helped analysing the power and limitations of different emerged structures (i.e. neural architectures). Feed-forward networks were able to make detour behaviour, but only recurrent networks (with Elman memory units) produced optimal solutions (for details see [21]). Because of space limitations, here we only want to stress how we "obliged" Khepera to learn the task. In a first series of experiments, we put Khepera directly in an environment containing a U-shaped barrier (such as in figure 5b). In this condition and with different neural architectures, Khepera was not able to "learn" the task. Then, we adopted a strategy to make a form of incremental learning.

For the first 200 generations, Khepera "lived" in an environment that contained a rectangular obstacle (a bar of 20 cm of length and 3 cm of height) and a target (a cylinder of 2 cm of diameter, see figure 5a). Each trial lasted a total of 240 input/output cycles divided into two epochs of 120 cycles each. At the beginning of each epoch, the Khepera was positioned in the environment behind the obstacle

facing North. The target was positioned beyond the obstacle once on the left side
and once on the right side of the upper portion of the environment in such a way
that the target fell outside the organism's restricted visual field and therefore it
could not be perceived by the robot. Khepera had 120 cycles available to make a
detour around the obstacle and reach the target.

After 200 generations, the environment changed. The obstacle now had an in-
verted U-shape and was moved slightly toward North (see figure 5b). The robot
was placed behind the obstacle facing North at the beginning of each epoch as
in the preceding environment. Each trial lasted 6 epochs (of 120 cycles each). In
3 of the 6 epochs there was no obstacle. In the other 3 epochs, the target was
located once on the left side, once in the right side, and once in the middle of the
upper portion of the environment. This environment management (notice that we
only changed the environmental characteristics and not the fitness formula) drove
Khepera to perform the detour task in a similar way to the chicks.

In short, it seems possible to reach complex behaviours with a careful design of
the evolutionary environmental conditions. However, this practise needs more theoretical investigations. A first step in that direction is represented in [2].

3.3 Evolving Robot Morphology

Recently, there has been an increasing interest in evolvable hardware (EHW), but
the concept does not seem yet to be well-defined. There is a lot of discussion
within the EHW community on whether to evolve at gate-level or function-level,
use genetic programming or cellular programming, do online evolution or off-line
evolution. In our view, the best definition of EHW that includes all of the above,
is the one by Yao and Higuchi [36], who defines EHW to be hardware that can
change its architecture and behaviour dynamically and autonomously with its environment, and they advocate that "EHW should be regarded as an evolutionary
approach to behaviour design rather than hardware design" [36].

A great number of experiments have been performed in EHW, in which the emphasis has been to show the validity of evolving electronic circuits [6–9, 31]. A quite
different approach to EHW, namely an evolvable electro-biochemical system, has
been suggested by Kitano [11]. There has been some initial approaches using EHW
in robotics tasks, most notably by Thompson [33], and if one looks at the definitions of autonomous robots and of EHW, they appear indistinguishable: hardware
that can change its architecture and behaviour dynamically and autonomously
with its environment.

The traditional applications of EHW in robotics only evolved the control circuit architecture of the robots. However, the circuit architecture is only a part of
the hardware system, and ideally we would like to evolve the whole system. The
hardware of a robot consists of both the circuit, on which the control system is
implemented, and the sensors, motors, and physical structure of the robot.

We [15] have previously argued that *True EHW* should evolve the whole hardware
system, since the evolution and performance of the electronic hardware is largely
dependent on the other parts of the hardware that constitute the system. The
latter part is what we called the *robot body plan*. A robot body plan is a specifica-

tion of the body parameters. For a mobile robot, it might be types, number and position of sensors, body size, wheel radius, wheel base, and motor time constant. With one specific motor time constant, the ideal control circuit should evolve to a different control than with another motor time constant; different sensors demand different control mechanisms; and so forth. Further, the robot body plan should adapt to the task that we want the evolved robot to solve. An obstacle avoidance behaviour might be obtained with a small body size, while a large body size might be advantageous in a box-pushing experiment; a small wheel base might be desirable for a fast-turning robot, while a large wheel base is preferable when we want to evolve a robot that turns slowly; and so forth. Hence, the performance of an evolved hardware circuit is decided by the other hardware parameters. When these parameters are fixed, the circuit is evolved to adapt to those fixed parameters that, however, might be inappropriate for the given task. Therefore, in *true EHW*, all hardware parameters should co-evolve.

The definition of *true EHW* is in accordance with the above mentioned definitions of EHW and autonomous robots that classify such systems as being able to change their architectures and behaviours dynamically and autonomously with their environment. Indeed, the design of a complete autonomous system must include self-organisation, and self-organisation means adaptation of *both* morphology and control architecture.

Together with Lee and Hallam, we [13, 15] therefore investigated the possibility of co-evolution between robot controllers and robot body plans in simulation, and we are currently building a LEGO robot system that allows evolution of robot body plans for real robots. The experiments showed that controllers and robot body plans are tightly coupled and that the robot body plan is determining for the performance of a controller.

We also went a step further and constructed specific hardware pieces, that would allow us to study the evolution of the morphology on a physical robot. We built a set of ears with hardware programmable pre-amplifiers, delays and mixers for the Khepera robot [19]. The hardware allows us to approximate the auditory morphology of various crickets by adjusting the programmable delays and the summing gains. With the reconfigurable hardware, we can investigate the relationship between the auditory morphology, the con-specific song, and the internal control system that generates the phonotaxis behaviour shown by the female cricket in response to the call of a mate. One possible investigation is then to co-evolve controller and auditory morphology to give good phonotaxis to a specific song while giving good discrimination between different kinds of songs.

4 Towards Breeding Robots

In most evolutionary robotics studies, there is the underlying assumption that it is in some sense trivial to design the mathematical fitness functions that can be used by the evolutionary algorithm to guide the development of robot controllers for task achieving behaviours. Yet, in some cases it might be difficult to think of the right fitness function before doing empirical testing. Recent experiments (see above section 3.2 and [25, 27]) do indeed suggest that some kind of shaping is necessary in the process of evolving task achieving robot behaviours, and other important results regarding shaping is found in other robotic directions (see e.g. [3]). The difficulties in describing adequate fitness functions (based on extensive studies of the problem domain), are seen by us as one of the main reasons why practically no researchers in the evolutionary robotics field have yet achieved truly complex robot behaviours.

Thanks a LEGO research and development contract, we explored the possibility to allow children to develop robot controllers for their individual "needs" or "taste". Obviously, we could not use the traditional evolutionary robotics approach, since it would require the children to have knowledge about how to design mathematical fitness functions.

We also wanted an approach different from the traditional *LOGO programming of LEGO-robots [29, 30]. Between these two constraints (traditional programming vs. evolutionary robotics), we chose a third way. The general idea is an *Interactive Evolutionary Robotics* approach by which children can develop (or evolve) robot controllers in the simulator by choosing among different robot behaviours that are shown on the screen, and then, when they are satisfied with the simulated robot's behaviour, download the developed control system to the real LEGO robot and further play with it in the real environment.

The interactive evolutionary robotics approach is inspired by our previous work using interactive genetic algorithms to evolve simulated robot controllers, facial expressions and artistic images (see e.g. [26, 35]). In this approach, there is no need of programming knowledge, since all the end-user has to do is to choose between the solutions suggested graphically on the screen. Hence, there is no description of a fitness function, but the selection in the genetic algorithm is performed by the user.

Surprisingly, we observed that children, using our tool, have been able to produce most of the simple robot behaviours that have been developed by researchers in the evolutionary robotics field.

We started to build a couple of LEGO robots for testing purposes, one of which is shown in Figure 6 (for details see [17]). As mentioned above and extensively reported in [14, 16, 22], it is possible to build an accurate simulator that allows one to go from simulation to reality by basing the simulator on the robot's own samplings of sensor and motor responses. The LEGO simulator was built by this technique (described in section 2).

The architecture of the control system that calculates the motor output given a sensory input in the simulated and real LEGO robots was a feed-forward neural network that connected sensory input to motor output. The jeep-like LEGO robot had 8 input neurons (4 bumpers and 4 light sensors), and the output was coded in 6 output neurons. In order to obtain a controller that guides the LEGO robot to perform a desired task, one will have to choose the right values for the (in this case) 42 connection weights in the feed-forward neural network control system. In a sense, this corresponds to finding a specific point (or region) in a 42 dimensional space. This is, of course, impossible for children (and most adults). Therefore, we used an interactive genetic algorithm to develop these connection weights. This algorithm allowed children to develop the connection weights (i.e. the control sys-

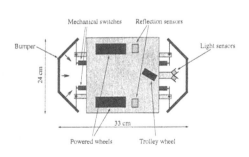

Figure6. Left: Schematic (upper) view of the jeep-like LEGO robot. The robot had three wheels: two powered wheels for locomotion and steering and one passive trolley wheel. The robot received sensory information from 4 mechanical switches mounted within two bumpers, 2 ambient light sensors at its front, and 2 reflection detectors under the chassis. Right: Photo of the LEGO robot. © Lund et al. 1997.

tem) by choosing among robot behaviours shown on the computer screen. The interactive genetic algorithm in the simulator works as follows.

The simulator allows different population sizes, but let us here look at a population of 9 robots. These 9 robots are placed in each their copy of the environment. The 9 similar environments are shown on the screen at the same time. Initially, the 9 robots in the population (generation 1), may have connections weights generated at random or loaded from a set of previously saved connection weights. The first of the 9 robots is put into the first environment. The robot will produce reactions based on the connection weights of the neural network control system and the sensory inputs. These reactions (movements) of the robot are shown in the first environment on the screen. After this robot has run around for a given time, the second robot is put into the (similar) second environment, and the produced movements are shown on the screen. This is done for all 9 individual robots. Hence, at this moment, the child has seen the 9 robots moving around in each their environment one after another. Let us look at the example where the robots' neural network control systems are generated at random. Some of the robots will differ in their behaviour, since the connection weights in the neural networks differ. The child might like some of these behaviours and dislike others. Based on such an evaluation where the child show preferences, the preferred robots in the population can be chosen to reproduce. The child can, for instance, choose 3 different ones, or choose the same robot 3 times.

In the reproduction phase, the 3 selected robots will be copied 3 times each in order to produce the next generation (generation 2) of 9 robots. However, as the neural network control systems are copied, mutation will be applied to 10% of the connection weights chosen at random. If a weight is selected, it will have added a value chosen at random in the interval [-10,10]. Hence, the new generation of robots will be similar to their parents because they are copies, but they will also differ to some degree because of the mutation.

The new generation of 9 robots is shown on the screen, one robot after another, as was the case with the previous generation. The child can then again select the 3 preferred robots and they will reproduce through copying and mutation to make the next generation (generation 3). This loop can continue until the child is satisfied with the behaviour of the LEGO robot. It is important to note, that the way the LEGO robots change is based on the child selecting which ones to reproduce according to the child's individual preferences.

Another way to change a LEGO robot's behaviour is to develop a single LEGO robot's behaviour rather than develop on a whole population. If this option is chosen, an enlarged image with the environment and robot will appear on the screen (see Figure 7 (right)). The robot will start moving around in this environment, just as it did when the whole population was presented. Now, if the child dislikes a movement of the robot (e.g., the robot gets stuck when hitting into a wall), the child can press a "Bad" button. This will immediately change the underlying neural network control system of that robot. But, in this case, the change does not happen at random. The system keeps a record of the last input to the robot, and it is the connections weights going out from the input neurons that were activated that are changed (with random portions). In the case where the robot hit into the wall with the front left bumper, it will be the connection weights going out from the input unit for front left bumper that are changed. Hence, the reaction upon a bump with the front left bumper is changed. Each time the "Bad" button is pressed, there is also a slight chance of the bias connections being changed, resulting in a change in the default behaviour.

When the child is satisfied with the behaviour obtained in the teaching process, the child can return to the whole population. The trained robot keeps the trained network, and can then, eventually, be selected to reproduce or to download on the real LEGO robot. In this way, we allow a Lamarckian evolution that consists of characteristics being learnt during life-time being inherited over evolution. We by no means support such a theory for natural evolution, but as a means for fast development of LEGO robots, it has proved to be a very powerful tool.

In a sense, what this type of reinforcement learning does is to make a guided search in specific regions. In the case with the bump with the front left bumpers, the algorithm performs a search on the weights connecting that input unit to the output units, and the search is made only on these weights. It seems to quickly find a solution, because the dimension of the search space has been limited. This is not a traditional reinforcement learning algorithm, but for the simple tasks of the LEGO robots we found that it works very well. We performed a series of experiments with the interactive evolutionary robotics approach described above, and had a lot of children playing with the LEGO robots. As expected, it turns out to be very simple to evolve exploratory behaviours, different kinds of obstacle avoidance, line & wall following, etc. and combinations between such simple behaviours. Hence, within minutes, children can develop a wide range of the robot behaviours that have traditionally been evolved in the field of evolutionary robotics. This is due to the use of the interactive evolutionary robotics approach combined with reinforcement learning. Indeed, with this approach children do not need to have

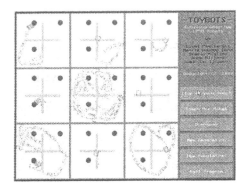 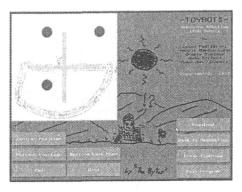

Figure7. Left: The population of 9 robots placed in each their (similar) environment. Here, the environment contains a couple of lines on the floor (the cross) and three round obstacles. In each environment, one can see the trajectory of the robot that has moved around in this environment. Right: The enlarged environment in which the child can train a single Toybot through reinforcement learning. The control system of the Toybot is changed every time the "Bad" button is pressed. © Lund et al. 1997.

any programming skills whatsoever! All that is demanded is that the child specify individual preferences in order to develop robots according to the individual taste.

5 Conclusions

In this paper, we have outlined part of our joint work in evolutionary robotics during the past years. We have moved from artificial life simulation work to real robot experiments and showed how this is possible with a carefully designed robot simulator. This allowed us to test animal behavioural hypotheses directly by evolving real, physical agents to perform in identical experimental set-ups. These experiments indicated that some kind of shaping would be necessary when evolving complex robot behaviours, and indeed we are now introducing *breeding robotics* that combines shaping and evolutionary robotics by using interactive genetic algorithms and reinforcement learning. Our future efforts will aim at showing how truly complex behaviours can emerge from breeding robots.

Acknowledgements

Materials and facilities were provided by the LEGO Group, the Danish National Centre for IT Research – CIT, the National Research Council, Rome, Italy, and the University of Edinburgh. Thanks to co-authors of the referenced papers (Billard, Denaro, Hallam, Ijspeert, Lee, Nafasi, Nolfi, Pagliarini, Parisi, Tascini, Taylor, Vucic, Webb).

References

1. K. Cheng. A purely geometric module in the rat's spatial representation. *Cognition*, 23:149–178, 1986.
2. A Clark and C. Thornton. Trading Spaces: Computation, Representation, and Limits of Uniformed Learning. *Behavioral and Brain Sciences*, 20:57–90, 1997.
3. M. Dorigo and M. Colombetti. Robot Shaping: Developing Autonomous Agents through Learning. *Artificial Intelligence*, 71(2):321–370, 1994.
4. D. Floreano and F. Mondada. Automatic Creation of an Autonomous Agent: Genetic Evolution of a Neural Network Driven Robot. In D. Cliff, P. Husbands, J. Meyer, and S. W. Wilson, editors, *From Animals to Animats 3: Proceedings of the Third International Conference on Simulation of Adaptive Behavior (SAB94)*, Cambridge, MA, 1994. MIT Press, Bradford Books.
5. C. R. Gallistel. *The organization of learning*. MIT Press, Cambridge, MA, 1990.
6. H. Hemmi, J. Mizoguchi, and K. Shimohara. Development and evolution of hardware behaviours. In R. Brooks and P. Maes, editors, *Proceedings of ALIFE IV*, Cambridge, MA, 1994. MIT Press.
7. T. Higuchi, H. Iba, and B. Manderick. Evolvable Hardware with Genetic Learning. In H. Kitano, editor, *Massively Parallel Artificial Intelligence*, MA, 1994. MIT Press.
8. T. Higuchi, M. Murakawa, M. Iwata, I. Kajitani, W. Liu, and M. Salami. Evolvable Hardware at Function Level. In *Proceedings of IEEE Fourth International Conference on Evolutionary Computation*, NJ, 1997. IEEE Press.
9. T. Higuchi, T. Niwa, T. Tanaka, H. Iba, H. de Garis, and T. Furuya. Evolving hardware with genetic learning: A first step towards building a Darwin machine. In J. Meyer, H. L. Roitblat, and S. W. Wilson, editors, *From Animals to Animats II: Proceedings of the Second International Conference on Simulation of Adaptive Behavior (SAB92)*, Cambridge, MA, 1992. MIT Press-Bradford Books.
10. N. Jakobi, P. Husbands, and I. Harvey. Noise and The Reality Gap: The Use of Simulation in Evolutionary Robotics. In *Proceedings of 3.rd European Conference on Artificial Life (ECAL'95)*. Springer-Verlag, 1995.
11. H. Kitano. Towards Evolvable Electro-Biochemical Systems. In C. Langton and T. Shimohara, editors, *Proceedings of Artificial Life V*, Cambridge, MA, 1997. MIT Press.
12. W. Koelher. *The Mentality of Apes*. Harcout Brace, New York, 1925.
13. W.-P. Lee, J. Hallam, and H. H. Lund. A Hybrid GP/GA Approach for Co-evolving Controllers and Robot Bodies to Achieve Fitness-Specified Tasks. In *Proceedings of IEEE Third International Conference on Evolutionary Computation*, NJ, 1996. IEEE Press.
14. H. H. Lund and J. Hallam. Sufficient Neurocontrollers can be Surprisingly Simple. Research Paper 824, Department of Artificial Intelligence, University of Edinburgh, 1996.
15. H. H. Lund, J. Hallam, and W.-P. Lee. Evolving Robot Morphology. In *Proceedings of IEEE Fourth International Conference on Evolutionary Computation*, NJ, 1997. IEEE Press. Invited paper.
16. H. H. Lund and O. Miglino. From Simulated to Real Robots. In *Proceedings of IEEE Third International Conference on Evolutionary Computation*, NJ, 1996. IEEE Press.
17. H. H. Lund, O. Miglino, L. Pagliarini, A. Billard, and A. Ijspeert. Evolutionary Robotics — A Children's Game. In *Proceedings of IEEE Fifth International Conference on Evolutionary Computation*, NJ, 1998. IEEE Press.
18. H. H. Lund and D. Parisi. Pre-adaptation in populations of neural networks evolving in a changing environment. *Artificial Life*, 2(2):179–197, 1996.

19. H. H. Lund, B. Webb, and J. Hallam. A Robot Attracted to the Cricket Species *Gryllus bimaculatus*. In P. Husbands and I. Harvey, editors, *Proceedings of Fourth European Conference on Artificial Life*, pages 246–255, Cambridge, MA, 1997. MIT Press, Bradford Books.

20. J. Margules and C. R. Gallistel. Heading in the rat: Determination by environmental shape. *Animal Learning and Behavior*, 16:404–410, 1988.

21. O. Miglino, D. Denaro, G. Tascini, and D. Parisi. Detour Behavior in Evolving Robots. 1998. In these proceedings.

22. O. Miglino, H. H. Lund, and S. Nolfi. Evolving Mobile Robots in Simulated and Real Environments. *Artificial Life*, 2(4):417–434, 1996.

23. O. Miglino, K. Nafasi, and C. Taylor. Selection for Wandering Behavior in a Small Robot. *Artificial Life*, 2(1), 1995.

24. O. Miglino, S. Nolfi, and D. Parisi. Discontinuity in evolution: how different levels of organization imply pre-adaptation. In R. K. Belew and M. Mitchell, editors, *Adaptive Individuals in Evolving Populations: Models and Algorithms*. Addison-Wesley, 1996.

25. S. Nolfi and D. Parisi. Evolving non-trivial behaviors on real robots: A garbage collecting robot. *Robotics and Autonomous Systems*, 1997. To appear.

26. L. Pagliarini, H. H. Lund, O. Miglino, and D. Parisi. Artificial Life: A New Way to Build Educational and Therapeutic Games. In C. Langton and K. Shimohara, editors, *Proceedings of Artificial Life V*, pages 152–156, Cambridge, MA, 1997. MIT Press.

27. S. Perkins and G. Hayes. Incremental acquisition of complex behaviour using structured evolution. In *Proceedings of the International Conference on Artificial Neural Networks and Genetic Algorithms '97*. Springer-Verlag, 1997. To appear.

28. L. Regolin, G. Vallortigara, and M. Zanforlin. Object and Spatial Representations in Detour Problems by Chicks. *Animal Behavior*, 48:1–5, 1994.

29. M. Resnick. MultiLogo: A Study of Children and Concurrent Programming. Master thesis, MIT, 1988.

30. M. Resnick. LEGO, Logo, and Life. In C. Langton, editor, *Artificial Life*, MA, 1989. Addison-Wesley.

31. M. Sipper, M. Goeke, D. Mange, A. Stauffer, E. Sanchez, and M. Tomassini. The Firefly Machine: Online Evolware. In *Proceedings of IEEE Fourth International Conference on Evolutionary Computation*, NJ, 1997. IEEE Press.

32. C. Thinus-Blanc. *Animal Spatial Cognition*. World Scientific, Singapore, 1996.

33. A. Thompson. Evolving electronic robot controllers that exploit hardware resources. In F. Moran, A. Moreno, J.J. Merelo, and P. Charon, editors, *Advances in Artificial Life: Proceedings of 3.rd European Conference on Artificial Life*, Heidelberg, 1995. Springer-Verlag.

34. G. Vallortigara and M. Zanforlin. Position Learning in Chicks. *Behavioral Proceedings*, 12:23–32, 1986.

35. V. Vucic and H. H. Lund. Self-Evolving Arts — Organisms versus Fetishes. *Muhely - Hungarian Journal of Modern Art*, 104:69–79, 1997.

36. X. Yao and T. Higuchi. Promises and Challenges of Evolvable Hardware. In *Proceedings of First International Conference on Evolvable Systems: from Biology to Hardware*, Heidelberg, 1996. Springer-Verlag.

Off-Line Model-Free and On-Line Model-Based Evolution for Tracking Navigation Using Evolvable Hardware

Didier Keymeulen[1], Masaya Iwata[1],
Kenji Konaka[2], Ryouhei Suzuki[2], Yasuo Kuniyoshi[1],
Tetsuya Higuchi[1]

[1] Electrotechnical Laboratory
Tsukuba, Ibaraki 305 Japan
[2] Logic Design Corp.
Mito, Ibaraki 305 Japan

Abstract. Recently there has been great interest in the idea that evolvable systems based on the principles of Artificial Life can be used to continuously and autonomously adapt the behavior of physically embedded systems such as mobile robots, plants and intelligent home devices. At the same time, we have seen the introduction of *evolvable hardware*(EHW): new integrated circuits that are able to adapt their hardware autonomously and almost continuously to changes in the environment [11]. This paper describes how a navigation system for a physical mobile robot can be evolved using a Boolean function approach implemented on evolvable hardware. The task of the mobile robot is to track a moving target represented by a colored ball, while avoiding obstacles during its motion. Our results show that a dynamic Boolean function approach is sufficient to produce this navigation behavior. Although the classical *model-free* evolution method is often infeasible in the real world due to the number of possible interactions with the environment, we demonstrate that a *model-based* evolution method can reduce the interactions with the real world by a factor of 250, thus allowing us to apply the evolution process *on-line* and to obtain an *adaptive* tracking-avoiding system, provided the implementation can be accelerated by the utilization of evolvable hardware.

1 Introduction

Robotics has, until recently, consisted of systems able to automate mostly simple, repetitive and large scale tasks. These robots, e.g. arm manipulators, are mostly programmed in a very explicit way and in a well-defined environment. However for mobile robot applications, the environment must be perceived via sensors and is usually not fully specified. This implies that a mobile robot must be able to learn to deal with an unknown and possibly changing environment.

In this paper we tackle the navigation task for a mobile robot which must reach a goal, from any given position in an environment while avoiding obstacles.

The robot is regarded as a *reactive system* described by a dynamic Boolean function which is represented by a disjunctive normal form. The dynamic Boolean function can easily be implemented by evolvable hardware and change with the environment.

Unfortunately, the classical *model-free* evolution method, where the robot behavior is learned by evolution without learning a model of the environment, is infeasible in the real world due to the number of required fitness evaluations in the real world (in the order of $1,000,000$) which may need several hours. To avoid this problem, most of the model-free evolution methods use *off-line evolution*, e.g., the robot behavior is trained off-line using the training data. The robot's behavior will be fixed after training. However, good training data are very difficult to obtain, especially for real-world environments where we have little prior knowledge about them. A simulated environment has often been used instead in training. This raises the issue of how close the simulated environment might be of the real one. The main objective of the off-line model-free evolution method is to find a robust behavior to maintain the robot performance in the real world despite the gap between the simulated and real world.

We have shown in this paper that a *model-based* evolution method can alleviate this problem significantly. In this method, a robot tries to build a model of the environment while learning by *on-line evolution* how to navigate in this environment. Such simultaneous learning of the environment and navigation within the environment can reduce the number of fitness evaluations in the real physical environment since the robot can use the environment it builds progressively. Our experiments have shown that we can reduce the number of fitness evaluations in the real environment by a factor of 250 for the navigation task we considered. Fitness evaluations can be done extremely fast in the internal environment because they are done at electronic speed in hardware.

The paper first defines the robot task and its environment. In section 3, it describes the reactive navigation system based on a Boolean function controller represented in its disjunctive normal form. In section 4 it presents the implementation of the evolution mechanism on the Evolvable Hardware. In section 5 it describes and compares the model-free and the model-based evolution methods implemented using Evolvable Hardware.

2   Robot Environment and Task

The shape of the robot is circular with a diameter of 25 cm (Fig.1). It has 10 infra-red sensors mapped into 6 Boolean variables indicating the presence of objects at a distance smaller than 30 *cm*. It is equipped with a bumping sensor to detect when the robot hits an obstacle and with two cameras able to identify and track a colored object (the ball in Fig. 1). The cameras return one of the 4 sectors, covering 90 degrees each, in which the target is located. The robot is driven by two independent motor wheels, one on either side. This allows the robot to perform 8 motions: 2 translations, 2 rotations and 4 combinations of rotation and translation. The robot is controlled by a PC mother board connected to two

Fig. 1. Real robot.

transputers: one dealing with the infra-red sensors, the vision sensor and the motor wheels, and the other controlling two EHWs which respectively, execute the robot behavior and simulate the evolutionary process. The environment is a world with low obstacles such that the colored target can always be detected by the robot. The obstacle shapes are such that using only a reactive system the robot will not become stuck. In other words, these are no overly complex shapes such horseshoes. For the off-line evolution approach, we built a "quasi exact" robot simulation to generate and evaluate the performance of the model-free method.

The task assigned to the robot is to reach the target without hitting obstacles within a minimum number of motions and from any position in the real world. To perform its task, the robot must learn two basic behaviors, obstacle avoidance and going to the target, and coordinate these two behaviors to avoid becoming stuck due to repetition of an identical sensor-motor sequence.

3 Reactive Navigation System

To describe our evolutionary approach, we consider a Markov decision process (MDP) model of robot-world interaction widely used in robot learning [35]. In this model the robot and the world are represented by two synchronized processes interacting in a discrete time cyclical process. At each time point, (i) the robot directly observes the world state, (ii) based on this current world state, the robot chooses a motion to perform, (iii) based on the current world state and the motion selected by the robot, the world makes a transition to a new state and generate a reward, and (iv) finally the reward is passed back to the robot.

One way to specify a robot's behavior is in terms of a controller, which prescribes, for each world state, a motion to perform. For the Markovian tracking-avoiding task we considered, a simple dynamic Boolean function control system will be used. The system can change or evolve its function with time. It assumes neither knowledge of the necessary behaviors nor the high level primitives of

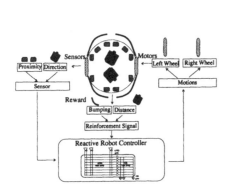

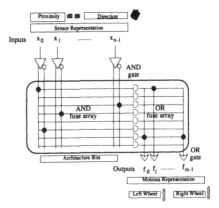

Fig. 2. Reactive navigation system.

Fig. 3. Evolvable Hardware Controller (EHW).

each behavior. It is well suited for an evolutionary search algorithm and is easily implementable in hardware. However to perform more complex tasks which are not Markovian such as navigation in an environment with obstacles of arbitrary shape (e.g., horseshoe shape) or where the target is not always visible, it may be necessary to exploit properties of the task, the sensor configurations, the environment and to change the existing control structure.

Formally, the controller is a function $\mathcal{F}$ from world states to motions. The Boolean function approach describes the function $\mathcal{F}$ as m Boolean functions of n Boolean variables which represents the desired reactive behavior. The input domain of the function $\mathcal{F}$ is $\{0,1\}^n$ where 2^n is the number of possible world states directly observable by the robot. It is encoded by 8 Boolean variables in our study: 6 bits for the infra-red sensors and by 2 bits for the vision sensor. It represents 256 world states observable by the robot. The output range of the function is $\{0,1\}^m$ where 2^m is the number of possible motions. It is encoded by 3 bits to represent 8 possible motions.

A possible representation of function $\mathcal{F}$ is a look-up table [17]. But the look-up table is generally impractical for real-world applications due to its space and time requirements and that it completely separates the information they have about one input situation without influencing what the robot will do in similar input situations. We chose to represent function $\mathcal{F}$ by m Boolean formula in *k-term* DNF[3] which consists of a disjunction of at most k terms ($k = 50$ in our study), each term being the conjunction of Boolean variables or their complement, such that function $\mathcal{F}$ can easily be implemented in LSI hardware which has a limited number of logic gates. The function controller represents function $\mathcal{F}$ by m Boolean functions f_i in their k-term DNF.

To increase the computation speed by one order of magnitude, function $\mathcal{F}$ described by m Boolean functions in their k-term DNF is executed with an

---

[3] f_i in k-term DNF: $f_i = (x_0 \wedge \cdots \wedge x_{n-1})_0 \vee \cdots \vee (x_0 \wedge \cdots \wedge x_{n-1})_{k-1}$

EHW. The EHW structure suited for the execution of function $\mathcal{F}$ consists of an AND-array and an OR-array (Fig.3). Each row in the AND array calculates the conjunction of the inputs connected to it, and each column of the OR-array calculates the disjunction of its inputs [18]. In our experimental set-up with $n = 8$ Boolean input sensors (6 bits for the infra-red sensors and 2 bits for the vision sensor) and 3 Boolean outputs, the AND-OR array has $2 * n + 3 = 19$ columns and needs a maximum number of $2^n = 256$ rows. However to force the Boolean function to generalize, the number of rows can be reduced by merging rows with the same output. In our experiments, we were able to reduce the number of rows to $k = 50$. In this way, on average, $\frac{256}{50} \simeq 5$ input states are mapped to the same output.

4 Evolvable Reactive Navigation System

From a machine learning perspective, the tracking-avoiding *control* task where the motions performed by the robot influence its future input situations [14] is an *associative delayed reinforcement* problem where the robot must learn the best motion for each world state from a *delayed* reward signal [29] [33]. For learning this simple robot task in *unknown* and *dynamic* environments, researchers have applied evolution-based learning algorithms [22] to low level control architecture such as LISP-like programming languages [19][4][26], finite state automata [32], production rules (classifier systems) [35] [6] [9], process network [28] and neural networks [7] [2][24][5][10][13] [12].

In our experiments, the learning task consists to find the function $\mathcal{F}$, mapping 256 inputs (world states) to 8 outputs (motions), in a search space of 8^{256} functions from a given set of observable input-output pairs and a reward signal. For learning pure, instantaneous Boolean functions of the inputs from a delayed reinforcement signal, we chose the evolutionary approach. It is better suited for dynamic environments and large search spaces when a large number of inputs and motions make $\mathcal{Q}$ learning impractical [33]. The evolutionary method works in a similar way as Kaelbling's Generate-and-Test algorithm [15]. The evolutionary algorithm performs a parallel search in the space of Boolean functions in a genetically inspired way. The algorithm is implemented in hardware where the 950 architecture bits of the EHW are regarded as the chromosome for the genetic algorithm [18]. We built specific mutation and cross-over operators for the k-term DNF representation to change the architecture bits and to reduce the learning time [18].

5 Evolvable Hardware Methods

To learn the tracking-avoiding task, the robot must interact with the real world environment (*on-line*) or its "quasi exact" simulation (*off-line*) to obtain information which can be processed to produce the desired controller. There are two ways to proceed to obtain the controller [16]:

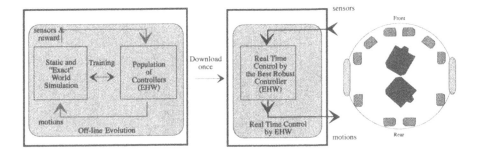

Fig. 4. Off-line Model-free evolution schema.

- **Model-free**: learn a controller without learning a model of the environment.
- **Model-based**: learn a model of the environment, and use it to derive a controller.

The next sections discuss and compare the model-free and model-based evolution. But for both methods, in order to give leverage to the learning process in a unknown and unpredictable real world we incorporate 3 biases:

- **reflexes**: We have programmed a reflex to find interesting parts in the environment. It moves the robot toward the ball when no obstacle is perceived around the robot.
- **shaping**: we present very simple environments to the robot first and then gradually exposes it to more complex environment.
- **local reinforcement signals**: we give a reinforcement signal that is local, helping the robot to step up a gradient.

5.1 Off-Line Model-Free Evolvable Hardware

The off-line model-free evolution simulates the evolutionary process of the robot controller in an artificial environment, simulating "quasi exactly" the real environment known a priori (Fig. 4). In this approach both the EHW and the environment are simulated to find the controller able to track a colored object and avoid the obstacles (left box in Fig. 4). Then the best controller found by evolution is used to control the real robot: the EHW architecture bits defining the best behavior are downloaded into the robot evolvable hardware board and control the robot in the real world (right box in Fig.4).

In this model-free approach, the evolutionary algorithm is used as an *optimization strategy* to find the optimal controller for a given simulated environment known a priori. The population size is 20 individuals. For the selection scheme, we have used a tournament selection with tournament size $s = 5$ and the elitist strategy to maintain the best individual. The main objective of the off-line evolution is to find a *robust controller* because the robot's controller cannot change

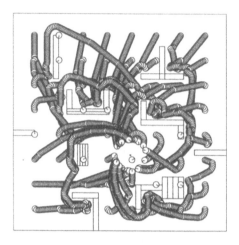

Fig. 5. Simulation of the motion of a real robot controlled by the best individual at generation 285 obtained by off-line model-free evolution.

during robot's life time. To obtain a robust controller, we first force the robot, during its evolution, to encounter many different situations. Second we improve the generalization ability of the controller by limiting the number of disjunctive terms in the k-term DNF representation.

Evaluation. Each robot is evaluated in the simulated environment. It fails when it cannot reach the target within a long period of time. It can fail for two reasons:

- It hits an obstacle.
- It reaches the maximum number of steps it is allowed to move in the environment. This situation occurs when the robot is stuck in a loop.

The fitness Φ of a controller is represented by a scalar between 0 (worst) and 64 (best) through combining three factors:

- $\mathcal{R}_1$: The number of times the robot has reached the target. When it reaches the target, it will be assigned to a new position in the environment. There are 64 new positions. This forces the robot to encounter many different world situations.
- $\mathcal{R}_2$: The distance to the target $D(robot, target)$, which forces the robot to reach the target. It is normalized using the dimension L of the simulated environment.
- $\mathcal{R}_3$: The number of steps used to reach its actual position from its initial position. This forces the robot to choose a shorter path and to avoid becoming stuck in a loop. An arbitrary large *Maximum Nbr. Steps* is used for normalization.

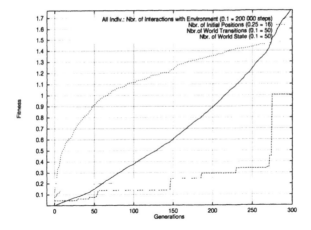

Fig. 6. Number of new positions of best individual (1 = 64 new positions) and interactions of all the individuals with the environment (1 = 2,000,000 interactions) throughout generations.

In our experiment the distance and the number of steps evaluations had an equal contribution to the fitness:

$$\Phi = \underbrace{Nbr.\ of\ New\ Positions}_{\mathcal{R}_1} + 0.5 \underbrace{\left(1 - \frac{D(robot, target)}{L}\right)}_{\mathcal{R}_2} +$$

$$0.5 \underbrace{\left(1 - \frac{Nbr.\ Steps}{Maximum\ Nbr.\ Steps}\right)}_{\mathcal{R}_3}$$

Experiments. We have conducted experiments with an environment containing 9 obstacles of different shapes except for horseshoes. The target is situated in the center of the environment (Fig.5).

During evolution and for each individual evaluation, the robot is always placed at the same initial position in the environment: the upper left corner. At the beginning of evolution, the controllers are initialized at random. The behavior of each individual is then simulated until it hits an obstacle or becomes stuck in a loop. When an individual reaches the target position, it is moved to a new position in the environment and its fitness is increased by 1. There are 64 new positions which are distributed equally in the environment and selected in a deterministic careful way introducing a shaping bias in the learning process. The important point is that during its evolution, all individuals start from the same initial position and that the sequence of new positions is always the same for all individuals.

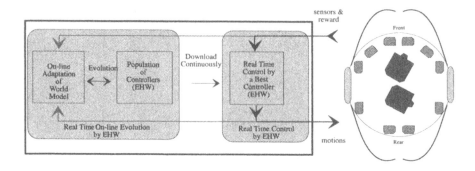

Fig. 7. On-line Model-based Evolution schema.

The behavior of the best individual in the population at generation 285 is shown in Figure 5. It demonstrates that the best individual coordinates the obstacle avoidance and the target tracking behaviors very well. For example, it discovers that a robust strategy is to turn clockwise around the obstacle until there is no obstacle in the direction of target.

Figure 6 shows the number of steps and the fitness of the best individual with generations. First and most importantly, figure 6 shows the number of interactions (nbr. of generations * nbr. of individuals * nbr. of steps) with the environment of all the individuals is $3,283,269$. It shows also that although there are 256 possible world states and 2048 possible transitions, the individuals in the population have encountered 122 world states and 744 transitions during their evolution. The number of genetic operations is 5700 (nbr. of individuals * nbr. of generations).

Once the best individual reaches the target 64 times, we download the controller of the best individual in generation 285 into the EHW. Although the real world differs from the simulated world in diverse aspects, the controller was robust enough to work well in the real world known a priori.

5.2 On-Line Model-Based Evolvable Hardware

The off-line model-free approach assumes that the simulated world is designed carefully and that the evolvable Boolean controller is robust enough to deal with the uncertainty and inaccuracy of the sensor values and the motors. Unfortunately, this approach cannot take into account any failures of the robot hardware and does not allow the controller to be adaptive, e.g. changing its structure while performing its task. But simply replacing training in the simulated world with training in the real world presents two problems. First, it decreases considerably the efficiency of the evolutionary approach due to the number of interaction with the environment needed (10 days if an interaction takes 0.25 second) to learn an effective behavior. Second, it doesn't maintain good on-line performance because the off-line evolutionary approach is only interested in the end-result.

Fig. 8. On-line Model-based Evolvable Hardware board with 2 EHW's.

To learn with fewer interactions with the physical environment and maintain good on-line performance, the robot can do some experiments in an "approximated" world model of the environment [21][3]. This approach is especially important in applications in which computation is considered to be cheap and real-world experience costly. It has been explored in reinforcement learning and extended to deal with unknown environments by learning the world model during robot's interaction with its environment [34] [30][31]. In the evolutionary approach, Grefenstette proposed to use a world model known a priori (a parameterized quantitative model) and to calibrate it during robot's life time [8][25].

On-line Model-based Learning. Our method learns a model continually through robot's life time and, at each step, the current model is used to compute an optimal controller (Fig. 7).

In order to keep the world model simple, adaptive and without bias, it is rough predictive model built using the experiences encountered by the robot [20]. It is represented by a deterministic state transition function where the states are the past world states observed by the robot, the transition are the motion generated by the robot and the outcome is the resulting world state observed by the robot. In the nondeterministic real world environment, each motion can have multiple possible outcomes. To model the environment we decide that the world model generate only the most recent outcome. There is one exception: the transitions which cause the robot to hit an obstacle are never erased from the world model. The world model is learned during the robot's life and is continuously changing to represent the probabilistic world in a deterministic way.

The on-line model-based approach works as follows: at each time step, the on-line model-based evolution simulates the evolutionary process of the robot controllers in the current world model of the environment to compute the optimal controller (left box on Fig. 7). After a few generations (around 10), the best controller found by evolution is downloaded to the EHW to control the robot in the real world (right box on Fig. 7). Moreover, while the robot executes the behavior of the best controller, the model is updated. In this approach the learning phase and motion phase are concurrent: while searching a new optimal controller, the robot, controlled by the last optimal controller, continue to track the ball and gather environmental data.

In this model-based approach, the evolutionary algorithm is used as an *adaptive strategy* [1] to find continuously an optimal controller for an approximated world model changing slightly but continuously. The population size is increased to 500 individuals to maintain diversity in the population. For the selection scheme, we have used a tournament selection with tournament size $s = 20$ and the elitist strategy.

In order to accelerate the entire evolutionary process by one order of magnitude, this process is implemented in a special-purpose hardware located including an evolvable hardware next to the evolvable hardware controlling the robot (Fig8). The special purpose hardware evaluates a population of controllers, implemented by an evolvable hardware, with the world model, implemented by a look-up table.

On-line Model-based Evolution. Each controller is evaluated in the world model using an *experience replay* strategy [20]. Although the learning and the execution phase may be concurrent, we use them in a sequence to simplify the comparison between the two methods. The learning phase starts when the robot fails to reach the target during the execution phase in the real world as described for the off-line model-free evolution approach.

The fitness Φ of each individual controller is obtained by testing the controller for each world state of the world model. Each world state of the world model is presented to the controller which returns the corresponding motion. Then the transition found in the world model predicts the next world state. The process continues until (i) no such world state exists in the actual world model, (ii) the transition causes the robot to hit an obstacle, (iii) this world state was already tested previously and (iv) an infinite loop is detected.

The fitness Φ is represented by a scalar between 0 (worst) and 1 (best) obtained by combining three factors:

- $\mathcal{R}_1$: The number of crashes when the robot hits an obstacle.
- $\mathcal{R}_2$: The number of infinite loops detected.
- $\mathcal{R}_3$: The total distance covered by the robot for each world state. For this measure, we know the distances covered by the real robot for each of the 8 motions and its *Max. Distance.*.

In our experiment, limiting the number of crashes and the number of infinite loops are the most important and has a greater contribution to the fitness than the distance:

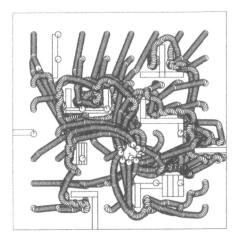

Fig. 9. Simulation of the motion of a real robot using a on-line model-based evolution and adapting continuously its controller.

$$\Phi = 0.5 \underbrace{\left(1 - \frac{Nbr.\,of\,Crash}{Nbr.\,of\,World\,States}\right)}_{\mathcal{R}_1} + 0.4 \underbrace{\left(1 - \frac{Nbr.\,of\,Infinite\,Loop}{Nbr.\,of\,World\,States}\right)}_{\mathcal{R}_2} +$$

$$0.1 \underbrace{\left(\frac{Distance\,Covered}{Max.\,Distance\,*\,Nbr.\,of\,World\,States}\right)}_{\mathcal{R}_3}$$

Using these three factors and the reflexes, the real robot is able to reach the target by searching and switching controllers in the population every time the world model changes.

Experiments. Although the method is dedicated to real world robots, we analyze the advantage of the on-line model-based evolution by simulating the real robot and conducting the same experiments in the same environment as that for the off-line evolution. The target is situated in the center of the environment (Fig.9).

At the beginning of evolution, all controllers (individuals) are initialized at random and the world model is empty. One of the controllers in the population is chosen at random to control the behavior of the robot. The real robot gathers data and builds the world model by memorizing the experiences, until it hits an obstacle or reaches a maximum number of steps. Then the evolution phase starts for a few generations (around 10) to find a controller which doesn't hit an obstacle or is not stuck in a loop. The execution and gathering process resumes

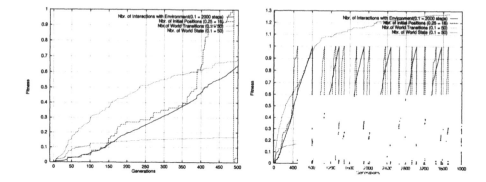

Fig. 10. Number of new positions of real robot (1 = 64 new positions) and of world interactions of real robot (1 = 20,000 interactions) throughout generations.

Fig. 11. Number of new positions (1 = 64 new positions) and of world interactions (1 = 20,000 interactions) executing 12 times 64 positions throughout generations.

with the new best controller after the evolution phase. When the robot reaches the target position, the robot is moved to a new position in the environment. There are 64 new positions and are identical as for the off-line evolution.

The behavior of the robot during its life time is shown in Figure 9. It demonstrates that the robot behavior changes its strategy during its life. For example, the way it follows a vertical obstacle changes: sometimes it strictly follows the wall other times it follows by bouncing on the wall.

Figure 10 shows the relation between the execution and evolution process by plotting the number of times the robot reached the target versus the number of generations. Also it shows the number of world state and transitions during the execution of the robot. It reached 84 world states and 334 transitions when the robot reached the target 64 times. But the world model was continuously changing, modifying the deterministic transition. The number of genetic operations is around 50 times larger than that for the model-free method: 242, 500 genetic operations. Finally and most importantly, it shows that the number of interactions of the robot with the environment is 12, 280, which is a factor of 250 smaller than the model-free evolution. It makes the model-based method using EHW, feasible in the real world.

Figure 11 is the extension of Figure 10 when we continue to assign new positions to the robot after 64 positions have already been assigned. It shows that although the robot needs less interactions (around 30 percent less) to reach the target 64 times, the on-line model-based evolution is a weak approach as an optimization strategy for a static environment because it is unable to find an optimal robust controller even after more than 120, 000 interactions and 3, 600 generations with the environment because its world model is approximated and continually changing.

The on-line model-based approach has been tested also in the real world for an environment with only one obstacle. The robot was able to avoid it in less than 5 *min* and to adapt if one of the sensor was blinded. These first experiments show that as a result of the on-line learning of world models and the evolvable hardware implementation, the number of interactions with the real world and the computation time to derive the controller can be reduced considerably. These results are encouraging. But many problems remain, such as how detailed the world model must be, whether it must include internal states and whether it must be probabilistic. All these questions are topics for future research.

6 Conclusion

We have demonstrated how EHW can be used to produce an on-line, model-based evolutionary navigation system for a mobile robot. The specific navigation task we addressed was the tracking of a colored target while avoiding obstacles. Our EHW produces completely reactive navigation control for this task by executing a Boolean function in its disjunctive normal form. Thus, we have demonstrated (1) that a reactive navigation system is able to perform the task of tracking and avoiding, (2) that the model-based approach allows us to build highly adaptive behaviors on-line, and (3) that a hardware implementation using EHW can maintain real-time robot performance.

Other tasks for which model-based on-line evolution with EHW is currently being investigated are data compression for ultra high precision images [27] and digital mobile communication [23]. Our research can be seen as part of this ongoing attempt to apply EHW to real world problems.

Acknowledgments

This research was supported by MITI Real World Computing Project (RWCP). The authors would like to express great thanks to Dr. Otsu and Dr. Ohmaki of ETL for their support and encouragement, Prof. Yao of New South Wales University for his invaluable comments and to Prof. Hoshino of Tsukuba University for his valuable discussions and his visionary thinking on artificial life.

References

1. Thomas Back, Ulrich Hammel, and Hans-Paul Schwefel. Evolutionary computation: Comments on the history and current state. *IEEE Transactions on Evolutionary Computation*, 1(1):3–18, 1997.
2. Randall D. Beer and J.C. Gallagher. Evolving dynamic neural networks for adaptive behavior. *Adaptive Behavior*, 1(1):91–122, July 1992.
3. Lashon B. Booker. Classifier systems that learn internal world models. *Machine Learning*, 3(2–3):161–192, 1988.
4. Rodney Brooks. Artificial life and real robots. In F. J. Varela and P. Bourgine, editors, *Proceedings of the First European Conference on Artificial Life*, pages 3–10, Cambridge, MA, 1992. MIT Press / Bradford Books.

5. Dave Cliff, Inman Harvey, and Philip Husbands. Explorations in evolutionary robotics. *Adaptive Behavior*, 2(1):73–110, July 1993.

6. Marco Dorigo and Marco Colombetti. Robot shaping: developing autonomous agents through learning. *Artificial Intelligence*, 71:321–370, 1994.

7. Dario Floreano and Francesco Mondada. Automatic creation of an autonomous agent: Genetic evolution of a neural-network driven robot. In J-A. Meyer Dave Cliff, Philip Husbands and S. Wilson, editors, *From Animals to Animats 3: Proceedings of the 3rd International Conference on Simulation of Adaptive Behavior*. MIT Press, 1994.

8. Johan J. Grefenstette and Connie Loggia Ramsey. An approach to anytime learning. In *Proceedings of the Ninth International Conference on Machine Learning*, pages 189–195, San Mateo, CA., 1992. Morgan Kaufman.

9. John J. Grefenstette and A. Schultz. An evolutionary approach to learning in robots. In *Proceedings of the Machine Learning Workshop on Robot Learning, Eleventh International Conference on Machine Learning*. New Brunswick, NJ, 1994.

10. Inman Harvey, Philip Husbands, and Dave Cliff. Seeing the light: Artificial evolution, real vision. In J-A. Meyer Dave Cliff, Philip Husbands and S. Wilson, editors, *From Animals to Animats 3: Proceedings of the 3rd International Conference on Simulation of Adaptive Behavior*. MIT Press, 1994.

11. Tetsuya Higuchi, Tatsuya Niwa, Toshio Tanaka, Hitoshi Iba, Hugo de Garis, and T. Furuya. Evolvable hardware with genetic learning: A first step towards building a darwin machine. In Jean-Arcady Meyer, Herbert L. Roitblat, and Stewart W. Wilson, editors, *Proceedings of the 2nd International Conference on the Simulation of Adaptive Behavior*, pages 417–424. MIT Press, 1992.

12. Tsutomu Hoshino, Daisuke Mitsumoto, and Tohru Nagano. Fractal fitness landscape and loss of robustness in evolutionary robot navigation. *Autonomous Robots*, 5:1–16, 1988.

13. Philip Husbands, Inman Harvey, Dave Cliff, and Geoffrey Miller. The use of genetic algorithms for the development of sensorimotor control systems. In F. Moran, A. Moreno, J.J. Merelo, and P. Chacon, editors, *Proceedings of the third European Conference on Artificial Life*, pages 110–121, Granada, Spain, 1995. Springer.

14. Leslie Pack Kaelbling. *Learning in Embedded Systems*. Bradford Book, MIT Press, Cambridge, 1993.

15. Leslie Pack Kaelbling. Associative reinforcement learning: A generate and test algorithm. *Machine Learning*, 15(3):299–320, 1994.

16. Leslie Pack Kaelbling and Andrew W. Moore. Reinforcement learning: A survey. *Journal of Artificial Intelligence Research*, 4:237–277, 1996.

17. Didier Keymeulen, Marc Durantez, Kenji Konaka, Yasuo Kuniyoshi, and Tetsuya Higuchi. An evolutionary robot navigation system using a gate-level evolvable hardware. In *Proceeding of the First International Conference on Evolvable Systems: from Biology to Hardware*, pages 195–210. Springer Verlag, 1996.

18. Didier Keymeulen, Masaya Iwata, Kenji Konaka, Yasuo Kuniyoshi, and Tetsuya Higuchi. Evolvable hardware: a robot navigation system testbed. *New Generation Computing*, 16(2), 1998.

19. John Koza. Evolution of subsumption using genetic programming. In F. J. Varela and P. Bourgine, editors, *Proceedings of the First European Conference on Artificial Life*, pages 3–10, Cambridge, MA, 1992. MIT Press / Bradford Books.

20. Long-Ji Lin. Self-improving reactive agents based on reinforcement learning, planning and teaching. *Machine Learning*, 8(3-4):297–321, 1992.

21. Sridhar Mahadevan. Enhancing transfer in reinforcement learning by building stochastic models of robot actions. In *Proceedings of the Ninth International Conference on Machine Learning*, pages 290–299, 1992.

22. Maja Mataric and Dave Cliff. Challenges in evolving controllers for physical robots. *Robotics and Autonomous Systems*, 19(1):67–83, November 1996.

23. Masahiro Murakawa, Syuji Yoshizawa, Isamu Kajitani, and Tetsuya Higuchi. Evolvable hardware for generalized neural networks. In Martha E. Pollack, editor, *Proc. of Fifteenth International Joint Conference on Artificial Intelligence*, pages 1146–1151. Morgan Kaufmann Publishers, 1997.

24. Domenico Parisi, Stefano Nolfi, and F. Cecconi. Learning, behavior and evolution. In *Proceedings of the First European Conference on Artificial Life*, pages 207–216, Cambridge, MA, 1992. MIT Press / Bradford Books.

25. Connie Loggia Ramsey and John J. Grefenstette. Case-based initialization of genetic algorithms. In *Proceedings of the fifth International Conference on Genetic Algorithms*, pages 84–91, San Mateo, CA., 1993. Morgan Kaufmann.

26. Craig W. Reynolds. An evolved, vision-based model of obstacle avoidance behavior. In *Artificial Life III*, Sciences of Complexity, Proc. Vol. XVII, pages 327–346. Addison-Wesley, 1994.

27. Mehrdad Salami, Masaya Iwata, and Tetsuya Higuchi. Lossless image compression by evolvable hardware. In *Proceedings of the Fourth European Conference on Artificial Life*, Complex Adaptive Systems, pages 407–416, Boston, USA, 1997. MIT Press / Bradford Book.

28. Luc Steels and Rodney Brooks, editors. *The Artificial Life Route to Artificial Intelligence: Building Embodied, Situated Agents*. Lawrence Erlbaum Assoc, 1995.

29. Richard S. Sutton. Learning to predict by the method of temporal differences. *Machine Learning*, 3(1):9–44, 1988.

30. Richard S. Sutton. Integrated architectures for learning, planning, and reacting based on approximating dynamic programming. In *Proceedings of the Seventh International Conference on Machine Learning*, pages 216–224, 1990.

31. Richard S. Sutton. Planning by incremental dynamic programming. In *Proceedings of the Eighth International Workshop on Machine Learning*, pages 353–357. Morgan Kaufmann, 1991.

32. Adrian Thompson. Evolving electronic robot controllers that exploit hardware resources. In F. Moran, A. Moreno, J.J. Merelo, and P. Chacon, editors, *Advances in Artificial Life: Proceedings 3rd European Conference on Artificial Life*, pages 640–656, Granada, Spain, 1995. Springer-Verlag.

33. Christopher J.C.H. Watkins and Peter Dayan. Q-learning. *Machine Learning*, 8(3):279–292, 1992.

34. Steven D. Whitehead and Dana H. Ballard. A role for anticipation in reactive systems that learn. In *Proceedings of the Sixth International Conference on Machine Learning*, pages 354–357, 1989.

35. Stewart Wilson. Classifier systems and the animat problem. *Machine Learning*, 2:199–228, 1987.

Incremental Evolution of Neural Controllers for Robust Obstacle-Avoidance in Khepera

Joël Chavas, Christophe Corne, Peter Horvai, Jérôme Kodjabachian, and
Jean-Arcady Meyer

AnimatLab, Ecole Normale Supérieure, France

Abstract. A two-stage incremental approach has been used to simulate the evolution of neural controllers for robust obstacle-avoidance in a Khepera robot and has proved to be more efficient than a competing direct approach. During a first evolutionary stage, obstacle-avoidance controllers in medium-light conditions have been generated. During a second evolutionary stage, controllers avoiding strongly-lighted regions, where the previously acquired obstacle-avoidance capacities would be impaired, have been obtained. The best controller thus evolved has been successfully downloaded on a Khepera robot. The SGOCE evolutionary paradigm that has been used in these experiments is described in the text. Additional experiments are required to assess the usefulness of the corresponding implementation details. Future research will target furthering the incremental evolutionary process and evolving more intricate behaviors.

1 Introduction

According to a recent review [24] of evolutionary approaches to neural control in mobile robots, it appears that the corresponding research efforts usually call upon a direct encoding scheme, where the phenotype of a given robot — *i.e.*, its neural controller and, occasionally, its body plan — is directly encoded into its genotype. However, it has often been argued (e.g., [12,16]) that indirect encoding schemes — where the genotype actually specifies developmental rules according to which complex neural networks and morphologies can be derived from simple programs — are more likely to scale up with the complexity of the control problems to be solved, if only because the size of the genotypic space to be explored may be much smaller than that of the space of the resultant phenotypes.

The feasibility of such indirect approaches, which combine the processes of evolution and development, has been demonstrated through several simulations [2,3,5,9,17,18,30,34,36,37] and a few applications involving real robots [6, 13,14,10,26,27]. However, the fact that the great majority of controllers and behaviors that have thus been generated are very simple, together with the difficulties encountered when more complex controllers and behaviors were sought [10,18], led us to suspect that so-called incremental approaches [4,11,21] should necessarily be used in conjunction with indirect encoding schemes in more realistic applications. In other words, according to such a strategy, appropriate controllers and behaviors should be evolved and developed through successive stages in which good solutions to a simpler version of a given problem are used iteratively to seed the initial population of solutions likely to solve a harder version of the same problem.

In [18] such an incremental strategy has been used to evolve and develop neural controllers that permitted a simulated insect to successively walk, follow an odor gradient, and avoid obstacles. In this paper, it is used within the context of an evolutionary robotics application, where neural controllers for robust obstacle-avoidance in a Khepera robot are automatically generated. This work calls upon a two-stage approach, in which controllers for obstacle-avoidance in medium-light conditions are first evolved, and then improved to operate also in more challenging strong-light conditions, when a lighted lamp is added into the environment. Comparisons with results obtained under the alternative one-shot strategy are provided and support the above-mentioned intuition about the usefulness of an incremental approach.

2 Material and methods

This section will describe the task to be accomplished, and the SGOCE[1] paradigm that underlies our methodology. This task derives from the characteristics and limitations of the sensory motor apparatus of Khepera, which will be briefly summarized hereafter. Likewise, a short description will be provided of how this sensory motor apparatus has been simulated in this work. As for the description of the SGOCE methodology, it will deal successively with the developmental code that links the genotype of the robot to its phenotype, the syntactic constraints that limit the complexity of the phenotypes generated, the evolutionary algorithm inspired from genetic programming [19, 20] that generates developmental programs, and the incremental strategy that helps produce neural control architectures likely to exhibit increasingly adaptive capacities.

2.1 The obstacle-avoidance task

The real Khepera. Khepera [29] is a circular-shaped miniature mobile robot — with a diameter of 55 mm, a height of 30 mm, and a weight of 70 g — that is mounted on two wheels and two small Teflon balls. In its basic configuration, it is equipped with eight proximity sensors — six on the front, two on the back — that may also act as visible-light detectors. The wheels are controlled by two DC motors with incremental encoders that move in both directions.

In each proximity sensor of Khepera, an infra-red light emitter and receiver are embedded. This hardware allows two things to be measured: the normal

---

[1] This name is the acronym for the expression "Simple Geometry Oriented Cellular Encoding".

ambient light — through receivers only — and the light reflected by the obstacles — using both emitters and receivers. In medium-light conditions, this hardware makes it possible to detect an obstacle a short distance away — not more than about 5 cm. However, under strong light conditions, the corresponding receptors tend to saturate : the light emitted by the robot and reflected by obstacles cannot be distinguished from ambient light and, thus, cannot be detected (Figure 1). Therefore, this work aims at automatically evolving a robust obstacle-avoidance controller likely to differentiate between the two light conditions and to take appropriate motor decisions.

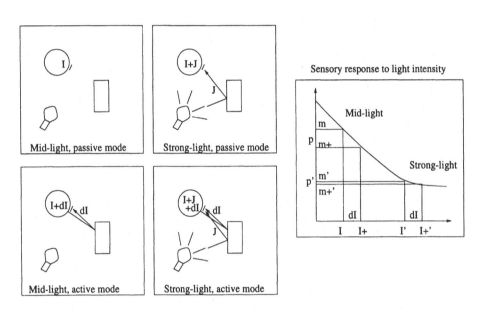

Fig. 1. (Left) This part of the figure shows the different sources that contribute to the intensity received by a sensor. The geometrical configuration considered is the same in all four situations. The ambient light contributes an intensity I. When the lamp is lighted, the sensor receives an additional contribution J, through direct and/or reflected rays. Finally, in active mode, the reflected IR-ray yields an intensity dI at the level of the sensor. The value of dI depends only on the geometrical configuration considered. Intensities are summed, but the sensor response is non-linear. (Right) In strong-light conditions, the response of a sensor can saturate. In such a case, where $I' = I + J$, the same increase in intensity dI caused by the reflection of an IR-ray emitted by the robot causes a smaller decrease of the value measured than it does in the linear region of the response curve (medium-light conditions). In other words, although dI intensity increases can be sensed in medium-light conditions — and the corresponding obstacle configurations can thus be detected —, such is not the case in strong-light conditions.

In the present work, such a controller has been generated through simulations performed under the SGOCE paradigm. Then the corresponding network has been downloaded onto a Khepera robot and its ability to generate the requested behavior has been checked.

The simulated Khepera. A cylindrical robot like Khepera is easier to model than a robot of arbitrary shape and with many degrees of freedom. Still, some phenomena, like friction, cannot be simulated with precision. Also, each sensor or motor has a unique behavior that can only be approximated in a simulation.

Our simulator is based on, and improves, an already existing simulator [25] and makes it possible to execute the same control program, either on the simulated robot or on the real one. It has four important features.

Firstly, it can be controlled by another independant program, making it easier to interface it with an already existing evolutionary algorithm software. This is important from a practical point of view, because the code can be reused more easily.

Secondly, it contains a set of functions specifically designed for artificial neural network evolution. One such function makes it possible for the evolutionary software to send to the simulator the description of a dynamic neural network, which will be connected in a specific way with the sensors and motors of the robot, whether real or simulated. Another function makes it possible to simulate the dynamics of that network during a given period of time, in order to control the robot. This function returns a fitness value, which is computed on the basis of information normally available to the robot, and which, when used with the real robot, is run entirely on board.

A third important feature of our simulator is its speed. Integer calculations are used to update the state of the neural network when computations are performed on board. Moreover, the sensor simulation method used by Michel has been replaced by a tabulation technique, according to which, prior to evolution, the values returned by a sensor in a given environment are recorded in a look-up-table for a number of different positions and orientations. Note that unlike in [28], where the values stored were measured on the real robot, here, we synthetize these values to make it easier to change the environmental conditions. At evaluation time, the sensor values are computed by interpolation from the values stored in the table.

Finally, another important feature to mention is the way in which sensor behavior is modelled. As already stated, Khepera IR sensors can work in either of two different modes. In passive mode, they return a measure m of the ambient light intensity I. In active mode, they return m^+, a measure of the intensity I^+, i.e., the sum of the ambient intensity I and of the intensity dI of the light possibly reflected off an obstacle (Figure 1).

If the robot is at a specific position relative to a given configuration of obstacles, then the value dI will be the same whatever the level I of the ambient light. The proximity measure $p = K \cdot (m - m^+)$ can thus be used to caracterize the presence of an obstacle. However, because the function relating intensity I to measure m is non-linear, the same dI value will not yield the same difference $(m - m^+)$ for different levels of I. For this reason, p is not simply a function of the intensity dI reflected from the IR-ray, but also depends on I.

We have modified Michel's IR sensor model in order to take into account the possible effect of the ambient light level I on p. At tabulation time, we sum the intensities conveyed by rays emitted by punctual light sources placed in the environment and possibly by the robot, which are received at the position of the sensor. Only then is the value of the measure returned by the sensor computed, using the response curve of Figure 1.

2.2 The SGOCE evolutionary paradigm

This paradigm is used to encode, into a robot's genotype, the developmental rules that will generate its phenotype. In the present application, this phenotype is instantiated as a general recurrent neural network controlling the behavior of the robot that is grown from a few initial cells provided by the experimenter. This neural network is made up of individual neurons each behaving as a leaky integrator [33] — i.e., as a universal dynamics approximator, liable to approximate the trajectory of any smooth dynamic system [1].

The developmental code. Our encoding scheme is a simple geometric variation of Gruau's cellular encoding [9]. It implements developmental rules that are encoded into artificial tree-like chromosomes that contain two categories of instructions. Some specify morphological transformations applying to specific cells, while others are used to generate structured developmental programs.

This scheme also employs a two-dimensional substrate within which the experimenter initially arranges a set of sensory cells that may be connected to the robot's sensors, a set of motoneurons that may be connected to the robot's actuators, and a set of precursor cells from which the developmental process will be initiated (Figure 2). Each precursor cell is given a copy of the robot's genotype and, as it executes the corresponding program, divides, grows connections to other cells, differentiates into a functional neuron, or dies (Figures 3 and 4).

At the end of such a process, a complete neural controller is obtained, whose architecture reflects the geometry and symmetries initially imposed by the experimenter, to a degree that depends on the side-effects of the developmental instructions that have been carried out. This controller is connected to the sensors and actuators of the robot through connections to the sensory cells and motoneurons incorporated into its architecture. This, together with the use of an appropriate fitness function (to be described later) makes it possible to assess the controller's capacity to generate the specific behavior sought by the experimenter.

Syntactic constraints. In order to reduce the size of the genotypic search-space and the complexity of the networks generated, we restrict the structure of the corresponding developmental programs by requiring that all evolving sub-programs be well-formed trees according to a given context-free tree-grammar (Figure 5). Furthermore, such a grammar makes it possible to control the nature and size of the program modifications that occur between two successive generations through use of genetic operators like mutation or crossover. Thus, a mutation operator makes it possible to replace a sub-tree by another randomly

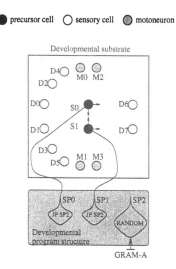

Fig. 2. An example of the initial state of the developmental substrate and of the structure of the developmental program. This structure is determined by the experimenter. Here, each of the two precursor cells carries out a developmental program that starts by a jump instruction to the evolving sub-program SP2. To solve other control problems, additional precursor cells might have been included in the substrate, each of which would execute a different sub-program before, occasionally, jumping to SP2 or to other evolving sub-programs. During evolution the composition of sub-program SP2 can be modified within the limits defined by the constraints encoded in grammar GRAM-A (see explanations on syntactic constraints in the text). $D0$ to $D7$ are sensory cells and $M0$ to $M3$ are motoneurons, which have been placed by the experimenter in specific positions within the substrate.

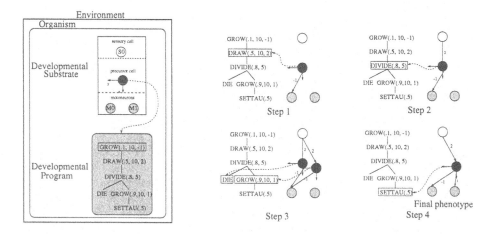

Fig. 3. While reading the genotype, a precursor cell generates a neural network after several developmental steps. This network may involve the sensory cells and the motoneurons made available by the experimenter. Each precursor cell is associated with a frame of reference that is inherited by its daughter cell when division occurs.

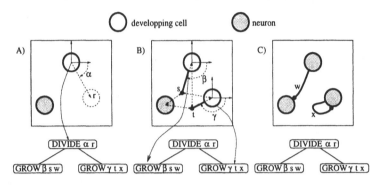

Fig. 4. Depending on the values of the arguments of some developmental instructions, targets for connections are sought in a given direction and at a given distance, in the local framework associated with the acting cell. These connections link two different neurons or correspond to self-connections. They can also regress and die when their targets lie outside the developmental substrate.

Terminal symbols
DIVIDE, GROW, DRAW, SETBIAS, SETTAU, DIE,
NOLINK, DEFBIAS, DEFTAU, SIMULT3, SIMULT4.
Variables
Start1, Level1, Neuron, Bias, Tau, Connex, Link.
Production rules
Start1⟶DIVIDE(Level1, Level1)
Level1⟶DIVIDE(Neuron, Neuron)
Neuron⟶SIMULT3(Bias, Tau, Connex) | DIE
Bias⟶SETBIAS | DEFBIAS
Tau⟶SETTAU | DEFTAU
Connex⟶SIMULT4(Link, Link, Link, Link)
Link⟶GROW | DRAW | NOLINK
Starting symbol
Start1.

Fig. 5. The GRAM-A grammar. This grammar defines a set of sub-programs — those that can be generated from it, starting with the *Start*1 symbol. When GRAM-A is used, a cell that executes such a sub-program undergoes two division cycles, yielding four daughter cells, which can either die or modify internal parameters (time-constant and bias) that will influence their future behavior as neurons. Finally, each surviving cell establishes a limited number of connections, either with another cell, or with the sensory cells and motoneurons that have been positioned by the experimenter in the developmental substrate.

generated compatible[2] subtree. Likewise, a crossing-over operator makes it possible to exchange a sub-tree in one developmental program for a compatible sub-tree in another developmental program.

Evolutionary algorithm. To slow down convergence by favoring the creation of ecological niches, we use a steady-state evolutionary algorithm that involves a population of N randomly-generated well-formed programs distributed over a circle and whose mode of operation is outlined in Figure 6.

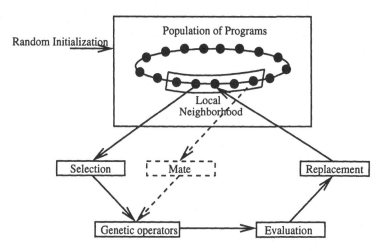

Fig. 6. The evolutionary algorithm (See text for explanation).

The following procedure is repeated until a given number of individuals have been generated and tested:

1. A position P is chosen on the circle.
2. A two-tournament selection scheme is applied in which the better of two programs randomly selected from the neighborhood of P is retained[3].
3. The program selected is allowed to reproduce, and three genetic operators may modify it. The recombination operator is applied with probability p_c. It exchanges two sub-trees between the program to be modified and another

---

[2] Two sub-trees are said to be compatible if they are derived from the same grammatical variable. For instance, if grammar GRAM-A (Figure 5) is used to define the constraints on a given sub-program, sub-tree SIMULT3(SETBIAS, DEFTAU, SIMULT4(GROW, DRAW, GROW, NOLINK)) in that sub-program may be replaced by sub-tree DIE, because both sub-trees are derivations of the *Neuron* variable in GRAM-A.

[3] A program's probability p_s of being selected decreases with the distance d to P: $p_s = max(R - d, 0)/R^2$, with R=4. Programs for which d is greater than or equal to R cannot be selected ($p_s = 0$).

program selected from the neighborhood of P. Two types of mutation are used. The first mutation operator is applied with probability p_m. It changes one randomly selected sub-tree into another randomly generated one. The second mutation operator is applied with a probability of 1 and modifies the values of a random number of parameters, implementing what Spencer called a *constant perturbation strategy* [35]. Firstly, the number n_{mut} of parameters to be modified is drawn from a binomial distribution $\mathcal{B}(n, p)$, and n_{mut} parameters are then selected randomly — all parameters having the same probability of being chosen — to be mutated.

4. The fitness of the new program is assessed by collecting statistics while the behavior of the animat controlled by the corresponding artificial neural network is simulated over a given period of time.

5. A two-tournament anti-selection scheme, in which the worst of two randomly chosen programs is selected, is used to decide which individual (in the neighborhood of P) will be replaced by the modified program.

In all the experiments reported in this paper, $N = 100$, $p_c = 0.6$, $p_m = 0.2$, $n = 6$ and $p = 0.5$.

Incremental approach. The artificial evolution of robust controllers for obstacle-avoidance was carried out using the Khepera simulator to solve successively two problems of increasing difficulty. Basically, this entailed evolving a first neural controller that used its sensors in active mode to measure the proximity value p, in order to avoid obstacles successfully in medium-light conditions. Then, a second neural controller was evolved that operated in passive mode in strong-light conditions and used measures of the ambient light level m to modulate the normal function of the first controller. In other words, such an incremental approach relied upon the hypothesis that the second controller would be able to evaluate the local intensity of ambient light so as to change nothing in the correct obstacle-avoidance behavior secured by the first controller in medium-lighted regions, but to alter it — in whatever adapted manner evolution would discover — when the robot travelled through strong-lighted regions likely to impair the proper operation of the first controller.

During Stage 1, to evolve the first controller and generate a classical obstacle-avoidance behavior in medium-light conditions, the following fitness function was used:

$$f_1 = \sum_t \left(0.5 + \frac{v_l(t) + v_r(t)}{4 \cdot V_{max}} \right) \cdot \left(1 - \frac{|v_l(t) - v_r(t)|}{2 \cdot V_{max}} \right) \cdot \left(1 - \frac{\sum_{front} p_i(t)}{4 \cdot P_{max}} \right) \quad (1)$$

where $v_l(t)$ and $v_r(t)$ were the velocities of the left and right wheels, respectively; V_{max} was the maximum absolute velocity; $p_i(t)$ was the proximity measure returned by each sensor i among the four front sensors; P_{max} was the largest measured value that can be returned.

In the righthand part of this equation, the first factor rewarded fast controllers, the second factor encouraged straight locomotion, and the third factor punished the robot each time it sensed an obstacle in front of it.

Using the substrate of Figure 2 and the grammar of Figure 5, controllers likely to include eight different sensory cells and four motoneurons were evolved after a random initialization of the population. The fitness of these controllers was assessed by letting them control the simulated robot over 500 time-steps, in a square environment containing an obstacle (Figure 7-a).

For this purpose, the sensory cells $D0$ to $D7$ in Figure 2 were connected to the robot's sensors such that the instantaneous activation value each cell propagated throught the neural network to the motoneurons was set to the proximity measure p returned by the corresponding sensor. Likewise, the motoneurons were connected to the robot's actuators such that a pair of motoneurons was associated with each wheel, the difference between their inputs determining the speed and direction of rotation of the corresponding wheel.

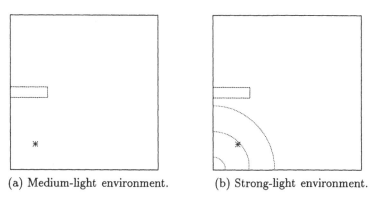

(a) Medium-light environment. (b) Strong-light environment.

Fig. 7. The environment that was used to evaluate the fitness of each controller. (a) Medium-light conditions of Stage 1. The star indicates the starting position of each run. (b) Strong-light conditions of Stage 2. A lighted lamp is positioned in the lower-left corner of the environment. Concentric arcs illustrate the corresponding intensity gradient.

Each controller was evaluated five times in the environment of Figure 7-a, starting in the same position, but with five different orientations, its final fitness being the mean of these five evaluations.

After 10,000 reproduction events, the controller with the highest fitness (called AVOID1 hereafter) has been used to seed an initial population that was subjected to a second evolutionary stage involving strong-light conditions. During Stage 2, the corresponding fitness function became:

$$f_2 = \sum_t \left(0.5 + \frac{v_l(t) + v_r(t)}{4 \cdot V_{max}}\right) \cdot \left(1 - \frac{|v_l(t) - v_r(t)|}{2 \cdot V_{max}}\right) \qquad (2)$$

In this equation, the third term that was included in the righthand part of equation 1 was eliminated because it referred to active-mode sensory inputs that could not be trusted in strong-light conditions. However, fast motion and straight locomotion were still encouraged.

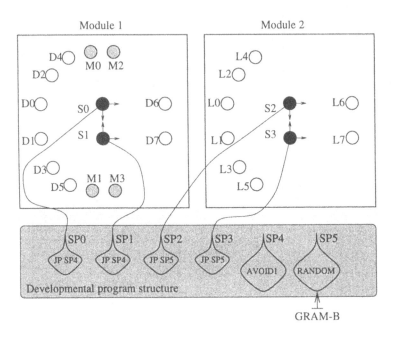

Fig. 8. The initial configuration of the developmental substrate and the program structure used for Stage 2. The same substrate is used for control experiments, in which both modules are evolved simultaneously. In these latter experiments, both SP4 and SP5 are initialized randomly and are submitted to evolution under constraints given by GRAM-A and GRAM-B, respectively.

Using the substrate of Figure 8 and the grammar of Figure 9, controllers likely to include 16 different sensors and four motoneurons were evolved during 10,000 additional reproduction events.

This time, the activation values that the new sensory cells $L0$ to $L7$ in Figure 8 propagated through a given controller were each set to the ambient light measure m returned by the robot's corresponding sensor. The fitness of the corresponding controller was assessed in the same square environment as the one used in Stage 1, but with a lighted lamp positioned in one of its corners (Figure 7-b).

Again, the final fitness was the mean of five evaluations that corresponded to five trials of 500 time-steps, each starting in the same position, but with different orientations and light intensities. In particular, one such trial was performed in medium-light conditions when the lamp was switched off, two others were performed when the lamp was contributing a small amount of additional light, and the last two were performed in strong light conditions, when the lamp contributed its maximum light intensity.

At the end of Stage 2, the best neural network thus obtained was downloaded and tested on a Khepera for 50 seconds.

Terminal symbols
DIVIDE, GROW, DRAW, GROW2, SETBIAS, SETTAU, DIE, NOLINK, DEFBIAS, DEFTAU, SIMULT3, SIMULT4.
Variables
Start1, Level1, Neuron, Bias, Tau, Connex, Link.
Production rules
Start1⟶DIVIDE(Level1, Level1)
Level1⟶DIVIDE(Neuron, Neuron)
Neuron⟶SIMULT3(Bias, Tau, Connex) | DIE
Bias⟶SETBIAS | DEFBIAS
Tau⟶SETTAU | DEFTAU
Connex⟶SIMULT4(Link, Link, Link, Link)
Link⟶GROW | DRAW | GROW2 | NOLINK
Starting symbol
Start1.

Fig. 9. The GRAM-B grammar. It is identical to GRAM-A except for the addition of one instruction (GROW2) that makes it possible to create a connection from the second to the first module.

3 Experimental results

The evolutionary run just described has been replicated ten times, the fitnesses of the best controllers obtained at the end of Stage 1 in medium-light conditions, on the one side, and at the end of Stage 2 in strong-light conditions, on the other side, being respectively given in columns 1 and 4 of Table 1.

Column 2 of Table 1 provides fitnesses that have been obtained when, at the end of Stage 1, each controller selected in medium-light conditions was transferred and tested in strong-light conditions. As for column 3 of Table 1, it provides fitnesses that were obtained when the best controllers selected at the end of Stage 2 were tested in medium-light conditions again.

The comparison of strong-light fitnesses indicates that controllers selected at the end of Stage 1 are less efficient than those that are obtained at the end of Stage 2 when the lighted lamp is added to the environment. Thus, this second evolutionary stage helped improving the behavior of the robot in strong-light conditions. Likewise, comparison of medium-light fitnesses indicates that the controllers selected at the end of Stage 2, according to their capacities at coping with strong-light conditions, didn't loss the essential of their abilities to generate appropriate behavior in medium-light conditions. Indeed, although their fitnesses

	Stage 1		Stage 2		Control	
run	medium-light	strong-light	medium-light	strong-light	medium-light	strong-light
1	323.253	097.104	345.746	359.786	245.934	258.740
2	326.283	102.576	296.758	309.380	240.169	252.095
3	339.701	159.331	171.105	224.239	241.174	252.125
4	302.774	105.761	280.408	314.011	397.838	183.006
5	321.416	101.187	301.514	345.597	254.706	226.549
6	396.320	107.621	258.273	239.041	228.097	230.399
7	204.426	230.999	193.847	235.449	245.495	246.534
8	312.598	185.041	296.922	262.145	203.891	232.949
9	310.051	219.790	255.324	304.069	170.998	224.549
10	398.471	181.863	346.630	409.280	221.416	231.129

Table 1. The performance of the best individuals of each run, when evaluated either in medium-light or in strong-light conditions. The incremental approach results in the performance values shown in columns 3 and 4 (Stage 2), while the control experiments results are shown in columns 5 and 6 (Control). The first two columns (Stage 1) provide results obtained after Stage 1 during the incremental approach.

tend to be slightly lower than those of the best controllers of Stage 1, they still are of the same order of magnitude.

To assess the usefulness of the incremental approach advocated here, ten additional control runs have been performed, each involving 10,000 reproduction events and the same number of evaluations (100,000) that were done in the incremental runs. Each such run directly started with the substrate of Figure 8 and called upon both GRAM-A and GRAM-B grammars, thus permitting the simultaneous evolution of both neural modules. The fitness of each individual was the mean of 10 evaluations: five in the conditions of Stage 1 described above, and five in the conditions of Stage 2. Columns 5 and 6 of Table 1 provide the fitnesses of the best individuals thus selected, these fitnesses having been assessed in both medium-light and strong-light conditions. A quantitative comparison of incremental and control runs indicates that the means of the medium-light and strong-light fitnesses obtained at the end of the incremental runs are statistically higher (Mann-Whitney test, significance level = 0.05) than the corresponding means obtained at the end of the control runs. Moreover, a qualitative comparison of the behaviors generated by these controllers indicate that the behaviors of the control runs are far less satisfactory than those of their incremental competitors. In fact, in every control run, but Run 4, the robot alternated moving forward and backward and never turned. As for the controller of Run 4, its fitness in medium-light conditions suddenly increased in the last generations and led to an obstacle-avoidance behavior as good as those of the best controllers evolved in the incremental runs, but at the detriment of its abilities to deal with strong-light conditions, which were severely impaired.

To understand how the controllers obtained during the incremental runs succeeded to generate satisfactory behaviors, the internal organization of the neural networks obtained at the end of Stage 1 (Figure 10) and Stage 2 (Figure 11) in one of these runs has been scrutinized. The corresponding simulated behaviors are respectively shown in Figures 12A-D. It thus appears that the single-module controller uses the four front sensors only. It drives the robot at maximal speed in open areas, and makes it possible to avoid obstacles in two different ways. If the obstacle is detected on one given side, then the opposite wheel slows down, allowing for direct avoidance. When the detection is as strong on both sides, then the robot slows down, reverses its direction, and turn slightly while recoiling. After a short period of time, forward locomotion resumes. However, when placed in strong-light conditions, this controller is unable to detect obstacles and thus keeps bumping into walls. The two-module controller corrects this default by avoiding strongly lighted regions in the following way. In medium-light conditions, all L-sensors return high m values. Excitatory links connect each of the two frontal L-sensors — $L0$ and $L1$ — to one interneuron of module 1 apiece. Each of these interneurons, in turn, sends a forward motion command to the motor on the opposite side. As a consequence, whenever the value m returned by one of these two sensors decreases — an event that corresponds to the detection of a high light intensity — the corresponding interneuron in Module 1 becomes less activated and the wheel on the opposite side slows down. This results in a light avoidance behavior.

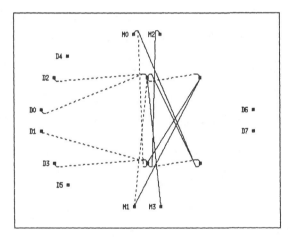

Fig. 10. The best controller obtained after Stage 1 for the particular run that resulted in the controller of Figure 11. The outputs of the motoneurons $M2$ and $M3$ are interpreted as forward motion commands for the right and left wheels, respectively, while the output of the motoneurons $M0$ and $M1$ correspond to backward motion commands. Solid lines correspond to excitatory connections, while dotted lines indicate inhibitory links. This network contains four interneurons.

Figures 13A-D show the behavior exhibited by Khepera when the networks described above are downloaded onto the robot and are allowed to control it for 50 seconds in a square arena of size 60x60 cm designed to scale the sim-

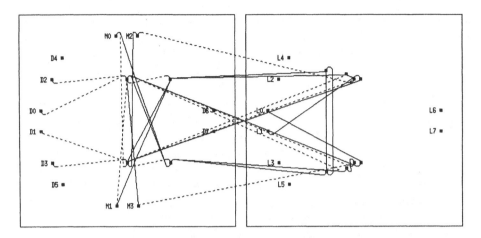

Fig. 11. The best controller obtained after Stage 2. This networks contains four interneurons in Module 1 and eight interneurons in Module 2.

ulated environment. Such figures were obtained through the on-line record of the successive orders sent to the robot's motors and through the off-line reconstruction of the corresponding trajectories. They demonstrate that the behavior actually exhibited by Khepera is qualitatively similar to the behavior obtained through simulation — in terms of the robot's ability to avoid obstacles and to quickly move along straight trajectories — and that it fits the experimenter's specifications. The main behavioral difference occurs when, at the end of the medium-light stage, controllers are tested in the presence of the additional lamp: the actual behavior of Khepera is more disrupted than the simulated behavior, probably because light that is reflected by the ground in the experimental arena is not adequately taken into account by the simulator (Figures 12.B and 13.B). Such discrepancies do not occur at the end of the strong-light stage because the robot then avoids the region where the lamp is situated and where such disturbing light reflections are the strongest.

4 Discussion

Although obstacle-avoidance would appear a behavior easy to evolve in Khepera, as demonstrated by the successful results already obtained by numerous researchers [6, 8, 7, 15, 22, 26–28, 32], results presented herein indicate that such a behavior is easily disrupted when the ambient light is high. These results also indicate that evolving a robust obstacle-avoidance behavior, although not trivial, is nevertheless possible. The solution that has been automatically discovered here consists in avoiding situations were the sensory capacities of the robot become too limited to secure a still adapted behavior. Finally, these results do not contradict the intuition that such non trivial behaviors are easier to evolve using a divide-and-conquer incremental approach. This intuition is further supported

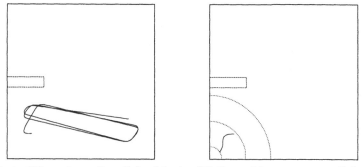

(A) Best of Stage 1, medium-light (sim.) (B) Best of Stage 1, strong-light (sim.)

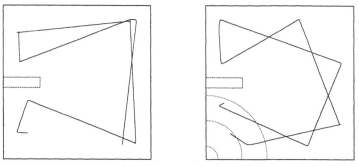

(C) Best of Stage 2, medium-light (sim.) (D) Best of Stage 2, strong-light (sim.)

Fig. 12. (Top) The simulated behavior of the single-module controller of Figure 10. When starting in front of the lamp, the corresponding robot gets stuck in the corner with the lamp. (Bottom) Simulated behavior of the two-module controller of Figure 11. Now, the robot avoids the lighted region of the environment.

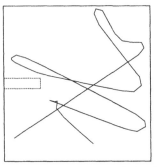

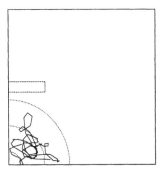

(A) Best of Stage 1, medium-light (real)   (B) Best of Stage 1, strong-light (real)

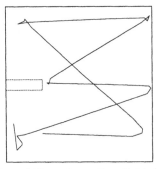

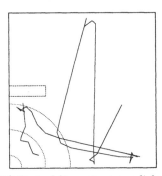

(C) Best of Stage 2, medium-light (real)   (D) Best of Stage 2, strong-light (real)

Fig. 13. The paths actually travelled by the Khepera robot, which have been reconstructed off-line using motor orders recorded during the trial. (Top) Real behavior of the single-module controller of Figure 10. Due to reflections present in the real world, but not in the simulation, the behavior under strong-light conditions is different from that of Figure 12. (Bottom) Real behavior of the two-module controller of Figure 11. Now, the real behavior is qualitatively similar to the simulated behavior shown in Figure 12.

by the observation that nature itself seems to resort to such an incremental approach, if one admits that, in an ever changing environment, selection pressures never remain constant, and if one observes that, since the appearance of life on Earth, the adaptive capacities of man clearly originate in the simpler adaptive capacities of numerous intermediate species.

Be that as it may, the results obtained herein are preliminary, and numerous additional experiments should be performed to assess the usefulness of a variety of implementation details. It seems *a priori* possible, for example, that individual neurons, behaving as traditional threshold units [23,31], (McClelland and Rumelhart, 1986; Rumelhart and McClelland, 1986), might be used to produce similar results to those obtained here with leaky integrators. We nevertheless used such neurons because their dynamic properties might prove to be mandatory in future extensions of this work.

Likewise, it is presently unclear whether the local interactions and the genetic operators that were used in our evolutionary algorithm are truly relevant and whether they might have been replaced by other options. Finally, one may wonder how integral each detail of the initial setup chosen by the experimenter — e.g., the grammars, the substrate's layout, the fitness functions — was for evolutionary success.

To further assess the potentialities of incremental evolution, future research efforts will aim at carrying on the evolutionary process one step further, resorting to a third evolutionary stage and a third fitness function. This might entail incorporating into Khepera's control architecture a rudimentary motivational system, according to which the robot — while still being able to avoid encountered obstacles — would seek the light when a simulated internal energy sensor detected low energy conditions, and would avoid light in normal or high energy conditions. Comparisons with a similar, but simpler, experiment carried on by [7] are likely to be enlightening because, in the latter approach, evolution was directly performed on the physical robot, i.e., without human intervention, and with a direct encoding scheme.

5   Conclusion

Preliminary results presented herein support the hypothesis that complex behaviors in real robots are more likely to be generated through an incremental evolutionary process than through direct evolution. They also suggest that realistic simulators may be devised, which permit neural controllers evolved in simulation to be successfully downloaded onto the corresponding robot. A two-stage incremental strategy made it possible to evolve a robust obstacle-avoidance behavior in a Khepera robot, although additional experiments are required to assess the relevance of each detail of the corresponding implementation. Future research will aim at elaborating the behavior thus far obtained through additional evolutionary stages that will manage a rudimentary motivational system.

References

1. Beer, R. D.: On the dynamics of small continuous-time recurrent neural networks. *Adaptive Behavior* **3(4)** (1995) 469–510
2. Boers, E. and Kuiper, H.: *Biological Metaphors and the Design of Modular Artificial Neural Networks.* Master's thesis, Dept. of CS and Exp. and The. Psy., Leiden University, (August 1992)
3. Cangelosi, A., Parisi, D. and Nolfi, S.: Cell division and migration in a 'genotype' for neural networks. *Network: computation in neural systems* (1995)
4. de Garis, H.: *Genetic Programming: GenNets, Artificial Nervous Systems, Artificial Embryos.* Ph.D. thesis, Université Libre de Bruxelles, Belgium, (1991)
5. Dellaert, F. and Beer, R.: Toward an evolvable model of development for autonomous agent synthesis. In R. A. Brooks, P. Maes, eds., *Proceedings of the Fourth International Workshop on Artificial Life.* The MIT Press/Bradford Books, Cambridge, MA, (1994)
6. Eggenberger, P.: Cell-cell interactions as a control tool of developmental processes for evolutionary robotics, (1996). Submitted to SAB96
7. Floreano, D. and Mondada, F.: Evolution of homing behavior in a real mobile robot. *IEEE Transactions on Systems, Man, and Cybernetics — Part B: Cybernetics* **26** (1996) 396–407
8. Floreano, D. and Mondada, F.: Automatic creation of an autonomous agent: Genetic evolution of a neural-network driven robot. In D. Cliff, P. Husbands, J.-A. Meyer, S. W. Wilson, eds., *From Animals to Animats 3. Proceedings of the Third International Conference on Simulation of Adaptive Behavior.* The MIT Press/Bradford Books, Cambridge, MA, (1994)
9. Gruau, F.: Automatic definition of modular neural networks. *Adaptive Behavior* **3(2)** (1994) 151–184
10. Gruau, F. and Quatramaran, K.: Cellular encoding for interactive evolutionary robotics. Tech. rep., University of Sussex, School of Cognitive Sciences, EASY Group, Brighton, UK, (1996)
11. Harvey, I., Husbands, P. and Cliff, D.: Seeing the light: Artificial evolution, real vision. In D. Cliff, P. Husbands, J.-A. Meyer, S. W. Wilson, eds., *From Animals to Animats 3. Proceedings of the Third International Conference on Simulation of Adaptive Behavior*, 392–401. The MIT Press/Bradford Books, Cambridge, MA, (1994)
12. Husbands, P., Harvey, I., Cliff, D. and Miller, G.: The use of genetic algorithms for the development of sensorimotor control systems. In P. Gaussier, J. Nicoud, eds., *From Perception to Action. Proceedings of the PerAc'94 Conference*, 110–121. IEEE Computer Society Press, Los Alamitos, CA, (1994)
13. Jakobi, N.: Evolutionary Robotics and the radical envelope of noise hypothesis. *Adaptive Behavior* **6(1)** (1997) 131–174
14. Jakobi, N.: Half-baked, ad-hoc and noisy: minimal simulation for Evolutionary Robotics. In Husbands, Harvey, eds., *Fourth European Conference on Artificial Life.* The MIT Press / Bradford Books, (1997)
15. Jakobi, N., Husbands, P. and Harvey, I.: Noise and the reality gap: The use of simulation in Evolutionary Robotics. In Moran, Moreno, Merelo, Chacon, eds., *Advances in Artificial Life: Proceedings of the Third European Conference on Artificial Life.* Springer Verlag, (1995)
16. Kodjabachian, J. and Meyer, J.-A.: Evolution and development of control architectures in animats. *Robotics and Autonomous Systems* **16(2–4)** (December 1995) 161–182

17. Kodjabachian, J. and Meyer, J.-A.: Evolution and development of modular control architectures for 1-d locomotion in six-legged animats, (1997). Submitted for publication in *Evolutionary Computation*

18. Kodjabachian, J. and Meyer, J.-A.: Evolution and development of neural networks controlling locomotion, gradient-following, and obstacle-avoidance in artificial insects, (1998). *IEEE Transaction on Neural Networks*, In press

19. Koza, J.: *Genetic Programming: On the Programming of Computers by Means of Natural Selection*. The MIT Press, (1992)

20. Koza, J.: *Genetic Programming II: Automatic Discovery of Reusable Subprograms*. The MIT Press, (1994)

21. Lee, W., Hallam, J. and Lund, H.: Proceedings of the 6th european workshop on learning robots. In *Learning complex robot behaviours by evolutionary approaches*. Brighton, (1997)

22. Lund, H. and Miglino, O.: From simulated to real robots. In *Proceedings of the 3rd IEEE International Conference on Evolutionary Computation*. IEEE Computer Society Press, (1996)

23. McClelland, J. L. and Rumelhart, D. E., eds.: *Parallel Distributed Processing*, vol. 1. The MIT Press/Bradford Books, Cambridge, MA, (1986)

24. Meyer, J.-A.: Evolutionary approaches to neural control in mobile robots. In *Proceedings of the IEEE International Conference on Systems, Man and Cybernetics*. San Diego, (October 1998)

25. Michel, O.: Une approche inspirée de la vie artificielle pour la synthèse d'agents autonomes, (1995). Submitted to EA95, Brest, France

26. Michel, O.: An artificial life approach for the synthesis of autonomous agents. In Alliot, Lutton, Ronald, Schoenauer, Snyers, eds., *Artificial Evolution*. Springer, (1996)

27. Michel, O. and Collard, P.: Artificial neurogenesis: An application to autonomous robotics. In Radle, ed., *Proceedings of the 8th. International Conference on Tools in Artificial Intelligence*. IEEE Computer Society Press, (1996)

28. Miglino, O., Lund, H. and Nolfi, S.: Evolving mobile robots in simulated and real environments. *Artificial Life* **2** (1995) 417–434

29. Mondada, F., Franzi, E. and Ienne, P.: Mobile robot miniaturization: A tool for investigation in control algorithms. In T. Yoshikawa, F. Miyazaki, eds., *Proceedings of the Third International Symposium on Experimental Robotics*, 501–513. Springer Verlag, Tokyo, (1993)

30. Nolfi, S., Miglino, O. and Parisi, D.: Phenotypic plasticity in evolving neural networks. In P. Gaussier, J. Nicoud, eds., *From Perception to Action. Proceedings of the PerAc'94 Conference*. IEEE Computer Society Press, Los Alamitos, CA, (1994)

31. Rumelhart, D. E. and McClelland, J. L., eds.: *Parallel Distributed Processing*, vol. 2. The MIT Press/Bradford Books, Cambridge, MA, (1986)

32. Salomon, R.: Increasing adaptativity through Evolutionary Strategies. In P. Maes, M. J. Mataric, J.-A. Meyer, J. B. Pollack, S. W. Wilson, eds., *From Animals to Animats 4. Proceedings of the Fourth International Conference on Simulation of Adaptive Behavior*. The MIT Press/Bradford Books, Cambridge, MA, (1996)

33. Segev, I.: Simple neuron models: Oversimple, complex and reduced. *Trends in Neurosciences* **15**(11) (1992) 414–421

34. Sims, K.: Evolving 3D morphology and behavior by competition. In R. A. Brooks, P. Maes, eds., *Proceedings of the Fourth International Workshop on Artificial Life*. The MIT Press/Bradford Books, Cambridge, MA, (1994)

35. Spencer, G.: Automatic generation of programs for crawling and walking. In K. E. K. Jr., ed., *Advances in Genetic Programming*, 335–353. The MIT Press / Bradford Books, Cambridge, MA, (1994)

36. Vaario, J.: *An Emergent Modeling Method for Artificial Neurol Networks*. Ph.D. thesis, University of Tokyo, (August 1993)

37. Vaario, J., Onitsuka, A. and Shimohara, K.: Formation of neural structures. In *Proceedings of the Fourth European Conference on Artificial Life, ECAL97*, 214–223. The MIT Press, (1997)

Springer
and the
environment

At Springer we firmly believe that an
international science publisher has a
special obligation to the environment,
and our corporate policies consistently
reflect this conviction.
We also expect our business partners –
paper mills, printers, packaging
manufacturers, etc. – to commit
themselves to using materials and
production processes that do not harm
the environment. The paper in this
book is made from low- or no-chlorine
pulp and is acid free, in conformance
with international standards for paper
permanency.

Lecture Notes in Computer Science

For information about Vols. 1–1397

please contact your bookseller or Springer-Verlag

Vol. 1398: C. Nédellec, C. Rouveirol (Eds.), Machine Learning: ECML-98. Proceedings, 1998. XII, 420 pages. 1998. (Subseries LNAI).

Vol. 1399: O. Etzion, S. Jajodia, S. Sripada (Eds.), Temporal Databases: Research and Practice. X, 429 pages. 1998.

Vol. 1400: M. Lenz, B. Bartsch-Spörl, H.-D. Burkhard, S. Wess (Eds.), Case-Based Reasoning Technology. XVIII, 405 pages. 1998. (Subseries LNAI).

Vol. 1401: P. Sloot, M. Bubak, B. Hertzberger (Eds.), High-Performance Computing and Networking. Proceedings, 1998. XX, 1309 pages. 1998.

Vol. 1402: W. Lamersdorf, M. Merz (Eds.), Trends in Distributed Systems for Electronic Commerce. Proceedings, 1998. XII, 255 pages. 1998.

Vol. 1403: K. Nyberg (Ed.), Advances in Cryptology – EUROCRYPT '98. Proceedings, 1998. X, 607 pages. 1998.

Vol. 1404: C. Freksa, C. Habel. K.F. Wender (Eds.), Spatial Cognition. VIII, 491 pages. 1998. (Subseries LNAI).

Vol. 1405: S.M. Embury, N.J. Fiddian, W.A. Gray, A.C. Jones (Eds.), Advances in Databases. Proceedings, 1998. XII, 183 pages. 1998.

Vol. 1406: H. Burkhardt, B. Neumann (Eds.), Computer Vision – ECCV'98. Vol. I. Proceedings, 1998. XVI, 927 pages. 1998.

Vol. 1408: E. Burke, M. Carter (Eds.), Practice and Theory of Automated Timetabling II. Proceedings, 1997. XII, 273 pages. 1998.

Vol. 1407: H. Burkhardt, B. Neumann (Eds.), Computer Vision – ECCV'98. Vol. II. Proceedings, 1998. XVI, 881 pages. 1998.

Vol. 1409: T. Schaub, The Automation of Reasoning with Incomplete Information. XI, 159 pages. 1998. (Subseries LNAI).

Vol. 1411: L. Asplund (Ed.), Reliable Software Technologies – Ada-Europe. Proceedings, 1998. XI, 297 pages. 1998.

Vol. 1412: R.E. Bixby, E.A. Boyd, R.Z. Ríos-Mercado (Eds.), Integer Programming and Combinatorial Optimization. Proceedings, 1998. IX, 437 pages. 1998.

Vol. 1413: B. Pernici, C. Thanos (Eds.), Advanced Information Systems Engineering. Proceedings, 1998. X, 423 pages. 1998.

Vol. 1414: M. Nielsen, W. Thomas (Eds.), Computer Science Logic. Selected Papers, 1997. VIII, 511 pages. 1998.

Vol. 1415: J. Mira, A.P. del Pobil, M.Ali (Eds.), Methodology and Tools in Knowledge-Based Systems. Vol. I. Proceedings, 1998. XXIV, 887 pages. 1998. (Subseries LNAI).

Vol. 1416: A.P. del Pobil, J. Mira, M.Ali (Eds.), Tasks and Methods in Applied Artificial Intelligence. Vol.II. Proceedings, 1998. XXIII, 943 pages. 1998. (Subseries LNAI).

Vol. 1417: S. Yalamanchili, J. Duato (Eds.), Parallel Computer Routing and Communication. Proceedings, 1997. XII, 309 pages. 1998.

Vol. 1418: R. Mercer, E. Neufeld (Eds.), Advances in Artificial Intelligence. Proceedings, 1998. XII, 467 pages. 1998. (Subseries LNAI).

Vol. 1419: G. Vigna (Ed.), Mobile Agents and Security. XII, 257 pages. 1998.

Vol. 1420: J. Desel, M. Silva (Eds.), Application and Theory of Petri Nets 1998. Proceedings, 1998. VIII, 385 pages. 1998.

Vol. 1421: C. Kirchner, H. Kirchner (Eds.), Automated Deduction – CADE-15. Proceedings, 1998. XIV, 443 pages. 1998. (Subseries LNAI).

Vol. 1422: J. Jeuring (Ed.), Mathematics of Program Construction. Proceedings, 1998. X, 383 pages. 1998.

Vol. 1423: J.P. Buhler (Ed.), Algorithmic Number Theory. Proceedings, 1998. X, 640 pages. 1998.

Vol. 1424: L. Polkowski, A. Skowron (Eds.), Rough Sets and Current Trends in Computing. Proceedings, 1998. XIII, 626 pages. 1998. (Subseries LNAI).

Vol. 1425: D. Hutchison, R. Schäfer (Eds.), Multimedia Applications, Services and Techniques – ECMAST'98. Proceedings, 1998. XVI, 532 pages. 1998.

Vol. 1427: A.J. Hu, M.Y. Vardi (Eds.), Computer Aided Verification. Proceedings, 1998. IX, 552 pages. 1998.

Vol. 1429: F. van der Linden (Ed.), Development and Evolution of Software Architectures for Product Families. Proceedings, 1998. IX, 258 pages. 1998.

Vol. 1430: S. Trigila, A. Mullery, M. Campolargo, H. Vanderstraeten, M. Mampaey (Eds.), Intelligence in Services and Networks: Technology for Ubiquitous Telecom Services. Proceedings, 1998. XII, 550 pages. 1998.

Vol. 1431: H. Imai, Y. Zheng (Eds.), Public Key Cryptography. Proceedings, 1998. XI, 263 pages. 1998.

Vol. 1432: S. Arnborg, L. Ivansson (Eds.), Algorithm Theory – SWAT '98. Proceedings, 1998. IX, 347 pages. 1998.

Vol. 1433: V. Honavar, G. Slutzki (Eds.), Grammatical Inference. Proceedings, 1998. X, 271 pages. 1998. (Subseries LNAI).

Vol. 1434: J.-C. Heudin (Ed.), Virtual Worlds. Proceedings, 1998. XII, 412 pages. 1998. (Subseries LNAI).

Vol. 1435: M. Klusch, G. Weiß (Eds.), Cooperative Information Agents II. Proceedings, 1998. IX, 307 pages. 1998. (Subseries LNAI).

Vol. 1436: D. Wood, S. Yu (Eds.), Automata Implementation. Proceedings, 1997. VIII, 253 pages. 1998.

Vol. 1437: S. Albayrak, F.J. Garijo (Eds.), Intelligent Agents for Telecommunication Applications. Proceedings, 1998. XII, 251 pages. 1998. (Subseries LNAI).

Vol. 1438: C. Boyd, E. Dawson (Eds.), Information Security and Privacy. Proceedings, 1998. XI, 423 pages. 1998.

Vol. 1439: B. Magnusson (Ed.), System Configuration Management. Proceedings, 1998. X, 207 pages. 1998.

Vol. 1441: W. Wobcke, M. Pagnucco, C. Zhang (Eds.), Agents and Multi-Agent Systems. Proceedings, 1997. XII, 241 pages. 1998. (Subseries LNAI).

Vol. 1442: A. Fiat. G.J. Woeginger (Eds.), Online Algorithms. XVIII, 436 pages. 1998.

Vol. 1443: K.G. Larsen, S. Skyum, G. Winskel (Eds.), Automata, Languages and Programming. Proceedings, 1998. XVI, 932 pages. 1998.

Vol. 1444: K. Jansen, J. Rolim (Eds.), Approximation Algorithms for Combinatorial Optimization. Proceedings, 1998. VIII, 201 pages. 1998.

Vol. 1445: E. Jul (Ed.), ECOOP'98 – Object-Oriented Programming. Proceedings, 1998. XII, 635 pages. 1998.

Vol. 1446: D. Page (Ed.), Inductive Logic Programming. Proceedings, 1998. VIII, 301 pages. 1998. (Subseries LNAI).

Vol. 1447: V.W. Porto, N. Saravanan, D. Waagen, A.E. Eiben (Eds.), Evolutionary Programming VII. Proceedings, 1998. XVI, 840 pages. 1998.

Vol. 1448: M. Farach-Colton (Ed.), Combinatorial Pattern Matching. Proceedings, 1998. VIII, 251 pages. 1998.

Vol. 1449: W.-L. Hsu, M.-Y. Kao (Eds.), Computing and Combinatorics. Proceedings, 1998. XII, 372 pages. 1998.

Vol. 1450: L. Brim, F. Gruska, J. Zlatuška (Eds.), Mathematical Foundations of Computer Science 1998. Proceedings, 1998. XVII, 846 pages. 1998.

Vol. 1451: A. Amin, D. Dori, P. Pudil, H. Freeman (Eds.), Advances in Pattern Recognition. Proceedings, 1998. XXI, 1048 pages. 1998.

Vol. 1452: B.P. Goettl, H.M. Halff, C.L. Redfield, V.J. Shute (Eds.), Intelligent Tutoring Systems. Proceedings, 1998. XIX, 629 pages. 1998.

Vol. 1453: M.-L. Mugnier, M. Chein (Eds.), Conceptual Structures: Theory, Tools and Applications. Proceedings, 1998. XIII, 439 pages. (Subseries LNAI).

Vol. 1454: I. Smith (Ed.), Artificial Intelligence in Structural Engineering. XI, 497 pages. 1998. (Subseries LNAI).

Vol. 1456: A. Drogoul, M. Tambe, T. Fukuda (Eds.), Collective Robotics. Proceedings, 1998. VII, 161 pages. 1998. (Subseries LNAI).

Vol. 1457: A. Ferreira, J. Rolim, H. Simon, S.-H. Teng (Eds.), Solving Irregularly Structured Problems in Prallel. Proceedings, 1998. X, 408 pages. 1998.

Vol. 1458: V.O. Mittal, H.A. Yanco, J. Aronis, R-. Simpson (Eds.), Assistive Technology in Artificial Intelligence. X, 273 pages. 1998. (Subseries LNAI).

Vol. 1459: D.G. Feitelson, L. Rudolph (Eds.), Job Scheduling Strategies for Parallel Processing. Proceedings, 1998. VII, 257 pages. 1998.

Vol. 1460: G. Quirchmayr, E. Schweighofer, T.J.M. Bench-Capon (Eds.), Database and Expert Systems Applications. Proceedings, 1998. XVI, 905 pages. 1998.

Vol. 1461: G. Bilardi, G.F. Italiano, A. Pietracaprina, G. Pucci (Eds.), Algorithms – ESA'98. Proceedings, 1998. XII, 516 pages. 1998.

Vol. 1462: H. Krawczyk (Ed.), Advances in Cryptology - CRYPTO '98. Proceedings, 1998. XII, 519 pages. 1998.

Vol. 1464: H.H.S. Ip, A.W.M. Smeulders (Eds.), Multimedia Information Analysis and Retrieval. Proceedings, 1998. VIII, 264 pages. 1998.

Vol. 1465: R. Hirschfeld (Ed.), Financial Cryptography. Proceedings, 1998. VIII, 311 pages. 1998.

Vol. 1466: D. Sangiorgi, R. de Simone (Eds.), CONCUR'98: Concurrency Theory. Proceedings, 1998. XI, 657 pages. 1998.

Vol. 1467: C. Clack, K. Hammond, T. Davie (Eds.), Implementation of Functional Languages. Proceedings, 1997. X, 375 pages. 1998.

Vol. 1468: P. Husbands, J.-A. Meyer (Eds.), Evolutionary Robotics. Proceedings, 1998. VIII, 247 pages. 1998.

Vol. 1469: R. Puigjaner, N.N. Savino, B. Serra (Eds.), Computer Performance Evaluation. Proceedings, 1998. XIII, 376 pages. 1998.

Vol. 1470: D. Pritchard, J. Reeve (Eds.), Euro-Par'98: Parallel Processing. Proceedings, 1998. XXII, 1157 pages. 1998.

Vol. 1471: J. Dix, L. Moniz Pereira, T.C. Przymusinski (Eds.), Logic Programming and Knowledge Representation. Proceedings, 1997. IX, 246 pages. 1998. (Subseries LNAI).

Vol. 1473: X. Leroy, A. Ohori (Eds.), Types in Compilation. Proceedings, 1998. VIII, 299 pages. 1998.

Vol. 1475: W. Litwin, T. Morzy, G. Vossen (Eds.), Advances in Databases and Information Systems. Proceedings, 1998. XIV, 369 pages. 1998.

Vol. 1477: K. Rothermel, F. Hohl (Eds.), Mobile Agents. Proceedings, 1998. VIII, 285 pages. 1998.

Vol. 1478: M. Sipper, D. Mange, A. Pérez-Uribe (Eds.), Evolvable Systems: From Biology to Hardware. Proceedings, 1998. IX, 382 pages. 1998.

Vol. 1479: J. Grundy, M. Newey (Eds.), Theorem Proving in Higher Order Logics. Proceedings, 1998. VIII, 497 pages. 1998.

Vol. 1480: F. Giunchiglia (Ed.), Artificial Intelligence: Methodology, Systems, and Applications. Proceedings, 1998. IX, 502 pages. 1998. (Subseries LNAI).

Vol. 1482: R.W. Hartenstein, A. Keevallik (Eds.), Field-Programmable Logic and Applications. Proceedings, 1998. XI, 533 pages. 1998.

Vol. 1483: T. Plagemann, V. Goebel (Eds.), Interactive Distributed Multimedia Systems and Telecommunication Services. Proceedings, 1998. XV, 326 pages. 1998.

Vol. 1487: V. Gruhn (Ed.), Software Process Technology. Proceedings, 1998. IX, 157 pages. 1998.

Vol. 1488: B. Smyth, P. Cunningham (Eds.), Advances in Case-Based Reasoning. Proceedings, 1998. XI, 482 pages. 1998. (Subseries LNAI).